Springer Proceedings in Physics 368

Indexed by Scopus

The series Springer Proceedings in Physics, founded in 1984, is devoted to timely reports of state-of-the-art developments in physics and related sciences. Typically based on material presented at conferences, workshops and similar scientific meetings, volumes published in this series will constitute a comprehensive up to date source of reference on a field or subfield of relevance in contemporary physics. Proposals must include the following:

- Name, place and date of the scientific meeting
- A link to the committees (local organization, international advisors etc.)
- Scientific description of the meeting
- List of invited/plenary speakers
- An estimate of the planned proceedings book parameters (number of pages/articles, requested number of bulk copies, submission deadline).

Please contact:

For Americas and Europe: Dr. Zachary Evenson; zachary.evenson@springer.com

For Asia, Australia and New Zealand: Dr. Loyola DSilva; loyola.dsilva@springer.com

Gagan B. Mohanty · Vaibhav S. Prabhudesai
Editors

Proceedings of the 2nd DAE Conclave—DAEC2026, 14–18 January, Mumbai, India

Innovative Solutions for a Sustainable Future Through Nuclear Horizons

Springer

Editors
Gagan B. Mohanty
Tata Institute of Fundamental Research (TIFR)
Mumbai, Maharashtra, India

Vaibhav S. Prabhudesai
Tata Institute of Fundamental Research (TIFR)
Mumbai, Maharashtra, India

ISSN 0930-8989 ISSN 1867-4941 (electronic)
Springer Proceedings in Physics
ISBN 978-981-92-2947-5 ISBN 978-981-92-2948-2 (eBook)
https://doi.org/10.1007/978-981-92-2948-2

This Springer imprint is published by the registered company Springer Nature Singapore Pte Ltd.
The registered company address is: 152 Beach Road, #21-01/04 Gateway East, Singapore 189721, Singapore

Foreword by Ajit Kumar Mohanty

The Department of Atomic Energy (DAE) has long pursued a dual mandate: advancing nuclear technology for energy and national security, while applying radiation technologies across agriculture, healthcare, industry, and research. It has also driven progress in fields like radio astronomy, supercomputing, particle accelerators, plasma science, and cryogenics. To strengthen collaboration across its institutions, DAE launched an annual conclave series beginning at NISER Bhubaneswar (2024) and followed by the second at TIFR Mumbai (Jan 14–18, 2026) during its Platinum Jubilee.

Apart from the scientific presentations highlighted in this proceedings, the conclave's highlight was a dynamic panel discussion on two nationally significant themes: Hydrogen production and storage, and Artificial Intelligence and Machine Learning (AI & ML). India's green hydrogen vision is strengthened by the Department's push for nuclear hydrogen, employing methods such as thermochemical cycles and electrolysis provided with nuclear heat as clean input from advanced reactors such as high-temperature gas-cooled reactors and molten salt reactors in the 5–10 MW range, which can operate at 800–900°C. DAE has already demonstrated laboratory-scale hydrogen production using the S-I cycle processes, and is setting up pilot-scale systems, with industrial deployment in view.

In parallel, the discussion on AI & ML highlighted their transformative impact across disciplines: medicine, physics, biology, and materials science, making the session both intellectually engaging and practically relevant. My assessment, shared by many participants at the conclave, is that the Department urgently needs large-scale centralized peta-flop computing facility, accessible to multiple research organizations.

Discussions on Accelerator-driven systems (ADS), detector development programs, Mega Science initiatives such as LIGO, MACE, GMRT and participation in international mega science project such as SKAO, development of a fourth-generation High Brilliance Synchrotron Radiation Source (HBSRS) with a 6 GeV electron storage ring at RRCAT, table top accelerator development, fabrication of qubits using superconducting circuits in alignment with the National Quantum mission's objectives, indigenous expertise by NFC to extract exceptionally pure niobium for superconducting cavities, and spearheading Thorium extraction from monazite, which also yields valuable rare-earth elements with wide spectrum applications provided insight on some of the programs pursued and the way forward.

It is a pleasure to mark the successful organization of this conclave. On behalf of the DAE fraternity, I extend sincere thanks to the Director of TIFR for graciously hosting the event. Prof. Gagan Mohanty, Prof. Vaibhav S. Prabhudesai, and the teams at TIFR and

DAE deserve special appreciation for their dedicated efforts in ensuring its success. The strong enthusiasm for the next conclave, scheduled even before this event concluded, reflects the impact. We now look forward to reconvening at NFC, Hyderabad, for the 3rd DAE Conclave.

Dr. Ajit Kumar Mohanty
Chairman, Atomic Energy Commission
Secretary, Department of Atomic Energy

Foreword by Vivek Bhasin

The Department of Atomic Energy (DAE) comprises a diverse ecosystem of R&D units, aided institutes, service organizations, and public sector undertakings, all contributing to the advancement of India's atomic and nuclear programmes in research, societal applications, and power generation.

DAE's recently articulated vision document organizes its key activities into major verticals, including reactors, nuclear fuels, advanced technologies, accelerators and lasers, basic science, societal applications, and human resource development. In this context, the DAE Conclave, an initiative of the Chairman, Atomic Energy Commission, Dr. Ajit Kumar Mohanty, serves as a vibrant platform to bring together scientists and engineers from across these domains to exchange ideas and strengthen collaborative execution.

Following the success of the inaugural edition in 2024, the second DAE Conclave was hosted by the Tata Institute of Fundamental Research, Mumbai, during January 14–18, 2026, under the theme *"Creating Innovative Solutions for a Sustainable Future through Nuclear Horizons."* The conclave provided a valuable forum to address emerging challenges in energy, sustainability, and technological innovation.

This volume brings together contributions spanning a wide spectrum of frontier areas. It highlights advances in reactor systems, hydrogen production using nuclear heat, and the closing of the nuclear fuel cycle, each critical to ensuring the nation's long-term energy security. It also reflects a renewed thrust in mega science and advanced technologies, including quantum computing and quantum communication. At its core, the volume remains anchored in the fundamental disciplines of science and technology—physics, chemistry, biology, and materials science.

I sincerely appreciate the organizers at TIFR and all contributors for making the conclave a success. The enthusiasm and intellectual rigor of the participants reflect the strength of the DAE community. As India moves towards *Viksit Bharat 2047*, nuclear science and technology will remain central, with BARC continuing to contribute with dedication, excellence, and a spirit of self-reliance.

Vivek Bhasin
Director, Bhabha Atomic Research Centre

Foreword by Jayaram Chengalur

It was a pleasure for TIFR to host the 2nd Department of Atomic Energy conclave "Creating Innovative Solutions for a Sustainable Future through Nuclear Horizons" from the 14th to the 18th of January 2026. Activities of the DAE span an enormous range, from all aspects of building and operating nuclear power plants, including the innovative R&D required for sustaining and progressing India's unique nuclear power ecosystem, as well as working towards futuristic technologies such as fusion power; applications of nuclear technology, including nuclear medicine and cancer treatment in general and in ensuring food security, as well as the fundamental research that is the bed rock on which a country's technological prowess is built.

In a system this large a diverse, it is easy for the staff working in one particular unit to not have a clear idea about all the other activities in the system. The DAE conclave was conceived of by the Secretary of the DAE, Dr. Ajit Kumar Mohanty, as an opportunity for the members of the DAE family to get a wholistic idea about all the activities that take place under the DAE umbrella. This event spread over several days also gives the scientists as well as the leaders and policy makers from across the department a chance to interact with each other and is an ideal forum for free discussions and the setting up of new collaborations.

The meeting itself had an exciting mix of talks and poster presentations spread over eight verticals, including such critical areas such as energy security, health care and quantum technologies. There were also stalls put up by the different units to showcase the activities of the unit. I thank all of the participants for the enthusiasm and energy that they brought to the conclave. I am delighted that the organizers are bringing out this volume with the written version of the contributions. This will serve not just as a record of the event, but also as an important reference for the future.

Jayaram Chengalur
Director, Tata Institute of Fundamental
Research

Preface

The Department of Atomic Energy (DAE) of the Government of India is mandated to develop and deploy nuclear technology for energy and national security, as well as to apply radiation technology in agriculture, health care, industry, and basic research. Since its inception in 1954, the department has made substantial contributions to several national and international programs. To celebrate its scientific achievements and strategize for the future, it has launched an annual series of conclaves, the second chapter of which (DAEC2026) was held at the Tata Institute of Fundamental Physics (TIFR) in Mumbai from January 14 to 18, 2026.

The conclave program featured presentations of key developments made across eight verticals: 1. Energy Security and Next-generation Reactors; 2. Quantum Technology; 3. Next-generation Materials; 4. Biology, Environment, and Healthcare; 5. Megascience Frontier; 6. Cutting-Edge Technology; 7. Mathematical Sciences; and 8. Science Education, Training, and Outreach. The event brought together about 1000 delegates, including scientists, technologists, researchers, and scholars from various DAE Units and Institutions, creating a structured platform for knowledge sharing, collaboration, and dialogue. A total of 205 posters were presented in two poster sessions, alongside 39 exhibit stalls that showcased the latest achievements of the Department's Units and Institutions. During DAEC2026, 39 talks were delivered, complemented by two panel discussions focused on Hydrogen Production and Storage Technology, and Artificial Intelligence and Machine Learning.

The 34 articles chosen for this volume of the *Springer Proceedings in Physics* highlight the glorious achievements of the DAE in recent years. The editors aim to convey the essence of these successes and leave their imprints in primary scientific literature.

Gagan B. Mohanty
Vaibhav S. Prabhudesai

Contents

Energy Security and Next Generation Reactors

Energy Security and Next Generation Nuclear Reactors

Vivek Bhasin(✉)

Bhabha Atomic Research Centre, Mumbai 400085, India
director@barc.gov.in

Abstract. India's nuclear energy programme has evolved as a cornerstone of the nation's long-term energy security, driven by a vision of self-reliance and sustainability. This article outlines the strategic role of research reactors developed at the Bhabha Atomic Research Centre in establishing a robust scientific and technological foundation for the country's nuclear journey. Beginning with early milestones such as APSARA and CIRUS, the programme enabled advancements in reactor physics, fuel cycle technologies, isotope production, and materials research, forming the backbone of India's three-stage nuclear power programme. The paper highlights the progression toward advanced systems, including high-flux research reactors, isotope production reactors, and next-generation concepts such as high-temperature gas-cooled reactors, molten salt reactors, and accelerator-driven systems. In the context of India's ambitious target of achieving ~ 100 GWe nuclear power capacity by 2047, the article discusses the expanding role of different types of reactors and closed fuel cycle strategies. Overall, the article underscores the critical contribution of sustained R&D, technological innovation, and institutional synergy within the Department of Atomic Energy in shaping a resilient, future-ready nuclear energy ecosystem aligned with the vision of a "Viksit Bharat."

Keywords: Closed Fuel Cycle · Plutonium and Thorium Fuel Cycle · High flux reactor · Nuclear Hydrogen Production · Radioisotopes and Radiopharmaceuticals · Accelerators Driven Subcritical System

1 Research Reactors at BARC: A Strategic Vision

At the formative stage of the Republic, India recognized the immense potential and futuristic possibilities of nuclear energy in ensuring the nation's long-term sustainable energy security. In a bold and visionary move, the Atomic Energy Commission was established in 1948, followed by the Department of Atomic Energy in 1954. India's nuclear energy programme was conceived at a time when the country's allied industrial base was still far from attaining the maturity required to substantially support such an ambitious endeavour. The journey of India's atomic energy programme has, therefore, been one of persistent challenges, ranging from periodic global isolation and technology-denial regimes to domestic fuel shortages. Yet, these very constraints became the driving force behind indigenization and self-reliance. Consequently, the Department of Atomic Energy (DAE) has built a long, proven, and proud legacy of harnessing atomic energy

G. B. Mohanty and V. S. Prabhudesai (Eds.): DAEC 2026, SPPHY 368, pp. 3–11, 2026.
https://doi.org/10.1007/978-981-92-2948-2_1

for peaceful purposes, including power generation and a wide spectrum of industrial and societal applications.

Homi Jehangir Bhabha, the architect of India's atomic energy programme, clearly understood that entry into the nuclear era required a strong and robust research foundation in nuclear science. Although nuclear research in India began even before Independence at Tata Institute of Fundamental Research under his leadership; the design, construction, and commissioning of India's nuclear research reactors (Fig. 1) commenced at Bhabha Atomic Research Centre, then known as the Atomic Energy Establishment, Trombay (AEET). The programme received a major impetus with the commissioning of Asia's first research reactor, the 1 MWth swimming-pool-type reactor APSARA, formally inaugurated by Jawaharlal Nehru in 1956. APSARA became the nucleus of an extensive research programme in nuclear science and technology, encompassing reactor design, nuclear fuels, materials science, nuclear electronics, and nuclear safety. It soon evolved into a core national facility for advanced research in neutron physics, fission physics, irradiation experiments for reactor technology, shielding studies, isotope production, radiation biology, and radiochemistry. The knowledge and experience gained through these efforts laid the backbone of India's future reactor programmes.

Recognizing the growing need for higher neutron flux, radioisotope production, and advanced irradiation facilities, India subsequently constructed the 40 MWth vertical tank-type reactor CIRUS, which was commissioned in 1960. CIRUS became the cornerstone of India's nuclear programme and served as the principal source of radioactive isotopes for many years. More importantly, it enabled the initiation of advanced neutron beam research, fuel and materials testing, and specialized training for nuclear scientists and engineers. It is noteworthy that the scientists and engineers of AEET indigenously produced the fuel rods for the reactor's first loading in 1960, exemplifying the spirit of technological self-reliance.

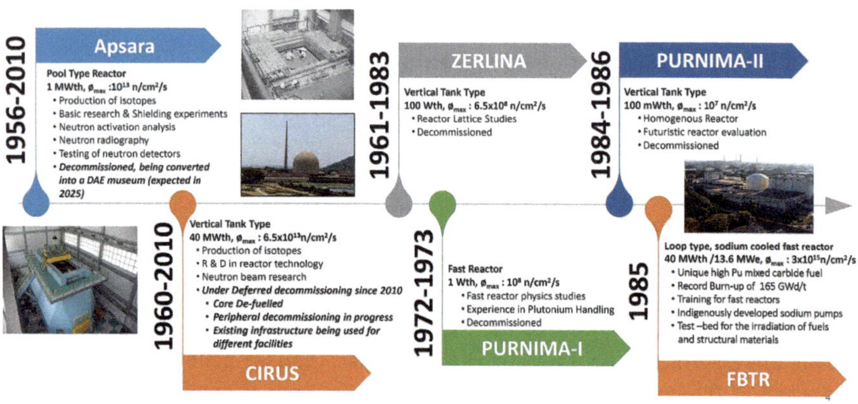

Fig. 1. Research rectors of India along with timeline

CIRUS also served as the cradle for R&D related to power reactors, including operation and maintenance practices, safety studies, and fuel-cycle research. The fuel element fabrication facility under the FAGGOT project was commissioned in 1961 for research reactors, which initially supplied fuel for RAPPS-1 as well. The plutonium recovered from reprocessing was later utilized in fuel fabrication for the Fast Breeder Test Reactor (FBTR). Subsequently, fuel used in FBTR at Kalpakkam was reprocessed at the COmpact facility for Reprocessing of Advanced fuels in Lead cells (CORAL) facility established in 2003 at Indira Gandhi Centre for Atomic Research, facilitating the second-stage recovery of plutonium for fast breeder reactor fuel fabrication. Importantly, CIRUS also enabled the early irradiation of thorium, producing the first fuel for India's thorium reactor programme. It is noteworthy that, in alignment with its three-stage nuclear strategy, India judiciously adopted Pressurized Heavy Water Reactors (PHWRs) as the cornerstone of its nuclear power generation programme from the outset.

ZERLINA served as a crucial experimental facility for studying lattice parameters of natural uranium-fuelled, heavy-water-moderated reactors and played a key role in optimizing PHWR design and validating reactor physics models. Beginning in the early 1970s and continuing through the 1990s, the PURNIMA series of reactors was developed for fundamental studies of plutonium and thorium fuel cycles. PURNIMA-I, India's first fast reactor critical assembly, attained criticality on May 18, 1972. These facilities made seminal contributions to the validation of nuclear data, reactor physics models, and core configurations, thereby advancing the country's fast reactor programme. India's growing self-reliance in nuclear science and engineering was further exemplified in 1985 with the indigenous design, development, and commissioning of DHRUVA, a 100 MWth tank-type research reactor fuelled with natural uranium and using heavy water as moderator, coolant, and reflector. Conceived to provide a high neutron flux for advanced research, isotope production, and materials testing, DHRUVA significantly strengthened India's nuclear power ambitions.

Over the decades, DAE has successfully developed and demonstrated the closed fuel cycle for Stage I at industrial scale, while the closed fuel cycle for Stage II has been successfully demonstrated at laboratory scale. In parallel, extensive R&D for the third stage was pursued through irradiation of thorium-bearing fuel bundles. The department has successfully demonstrated the recovery of U-233 and its deployment in reactors for thorium utilization, marking a significant milestone in India's long-term nuclear programme. The journey of engineering excellence continued with the commissioning of KAMINI at Kalpakkam in 1996. This 30 kWth tank-type reactor, fuelled with U-233–Al alloy plates produced from thorium irradiated in CIRUS, remains unique in the world. KAMINI serves as a neutron source for radiography, activation analysis, and detector calibration, while also providing an invaluable platform for studying U-233 fuel behaviour. Its continued operation underscores India's remarkable progress in thorium technology. Further advancing this trajectory, India commissioned a 100 Wth critical facility in 2008 to validate the physics design, safety studies, and performance parameters of future thorium-based reactors.

Most recently, India's nuclear research ecosystem was significantly strengthened with the commissioning of APSARA-U in 2018, the upgraded successor to APSARA. This 2 MWth pool-type reactor employs low-enriched uranium fuel and offers superior performance at higher burn-ups. Beyond neutron beam research, APSARA-U supports isotope production, neutron radiography, detector testing, and material characterization, while also serving as a premier training facility equipped with enhanced safety systems and advanced instrumentation.

2 Next Generation Research Reactors: Future Plan

2.1 High Flux Research Reactor (HFRR)

The HFRR with its high neutron flux to power ratio will be a leading research reactor having quality factor comparable to that of the advanced research reactors available in very few countries. Planned to be established at BARC-Vizag, this is a 40 MWth pool type reactor with maximum fast neutron flux of 1.0×10^{15} n/cm^2/s (~ five times higher than DHRUVA). The Reactor will use Low Enriched Uranium type fuel and demineralized water as coolant and moderator. It will be 'First of its Kind' facility in the country where 'Cold Neutron Source' will be available for advanced research related to materials and biology. In addition to fundamental research using neutron beam tube, and R&D support to nuclear power programme, it will also cater immensely in production of neutron transmutation doped (NTD) Silicon.

2.2 Isotope Production Reactor (IPR)

The IPR has been approved for construction at BARC-Vizag. The reactor is optimized for high-specific-activity isotope production, with advanced handling and processing systems. A dedicated IPR and co-located isotope processing facility (IPF) to be established for production of radioisotopes and radiopharmaceuticals to meet the domestic demand for making India self-reliant in the field of medical radioisotopes. The IPR is a 40 MWth tank-type reactor, cooled and moderated by light water and fuelled with low enriched uranium (LEU).

2.3 High Temperature Gas Cooled Reactor (HTGCR)

Hydrogen is one of the most promising options of future energy sources. Nuclear energy-based hydrogen production by splitting water is a sustainable and environmentally benign option for hydrogen production. High Temperature Reactor (HTR) technology development programme is aimed with a focus on operation at 650 °C, suitable for hydrogen production process. In this regard, a CO_2 cooled, graphite moderated up to 5 MWth High Temperature Gas-Cooled Reactor (HTGCR) is being developed which will be coupled with the chemical reactor to produce hydrogen. Essentially, it is a thermo-chemical generation of hydrogen, for which overall efficiency is very high. Since the hydrogen generation process will utilize nuclear heat, the land area requirement reduction is evident as compared to renewable sources of energy like solar, wind. Also, the energy equivalence of hydrogen would be cheaper than that of diesel and petrol for which the country faces import cost.

2.4 Molten Salt Reactor (MSR)

Research and development activities towards up to 5 MWth MSR is in progress. This is a second pathway for utilization of thorium, other than the classical three-stage program. MSR is an ambitious programme that aims to develop various new technologies in several areas such as high temperature material and components, fuel, high purity graphite etc. Experiments using proposed fuel salt has been initiated in BARC to calibrate the reactor physics codes, and results are quite encouraging.

2.5 Boron Neutron Capture Therapy (BNCT)

It is a low power reactor purely dedicated to Boron Neutron Capture Therapy (BNCT) and will be employed for treating the head and neck cancer by the alpha particle. The mechanism involves injection of boron compound into the patient, which gets accumulated in the malignant tissues, followed by bombardment of neutron generated in the reactor for transmutation of boron, which in turn releases alpha particle for killing of the cancer cells. This will be the first reactor, to be commissioned and operated in the public place, and is proposed to be established at Advance Centre for Treatment, Research and Education in Cancer (ACTREC) at Kharghar.

3 Nuclear Power Leading Energy Ecosystem for Viksit Bharat: Challenges and Mitigation

While nuclear research in the country continued to advance in sixties, in parallel the power programme also began with commissioning of twin units of Boiling Water Reactors (BWRs). Construction of PHWR started in collaboration with AECL of Canada and afterwards, a series of PHWRs set the tune of indigenization. Thus standardization, commercialization, and scaling-up of capacities from 220MWe to 540 MWe and now to 700 MWe have been achieved. The power programme has been expanded with 24 reactors with a capacity of 8780 MWe in the country. India now aspires to be "Viksit Bharat" by 2047 and envisages to enhance its nuclear energy capacity by 100 GWe by then. Necessary roadmap to reach this capacity has been drawn as shown in Table 1.

As evident, several new types of reactors have been envisioned in the future plan, especially small modular reactors (SMRs) for deployment as captive power plants for hard-to-abate industries, repurposing of retiring fossil fuel-based power plants, and deployment at off-grid, remote location. The Bharat Small Reactor (BSR) is actually India's first indigenized version with modification suitable as captive plant. The conceptual design is completed and no risk or uncertainty is foreseen for its success.

Following BSR, the Bharat Small Modular Reactor (BSMR) would be of immediate attention, and the initial activities including conceptual design has already been completed. Initially, a 220 MWe BSMR as demonstration plant has been planned to be commissioned at DAE site in close cooperation between BARC and NPCIL. Appreciable developments as far as the indigenous low alloy steel forgings and other important components have already been achieved, and equipment manufacturing is well within the capabilities of Indian industry. After successful demonstration of BSMR 220 MWe,

Table 1. Presently installed, progress, and future plan of nuclear power capacity of India

Particulars	Capacity (MWe)	Remarks
Present capacity	8780	Sites available/earmarked for DAE (~ 48 GWe)
Capacity under implementation + 1st Fleet (4 KKNPP + 13 PHWRs + PFBR)	13,600	
Future Capacity Additions		
2nd Fleet of PHWRs (2 × 700 MWe each) at existing sites of Kakrapar, Rajasthan, Narora and 4 PHWRs (4 × 700 MWe) at Bhimpur, MP approved in-principle	7000	
LWRs at Jaitapur (6 × 1730 MWe) & Kovvada (6 × 1208 MWe)	17,628	
2 × 500 MWe FBRs (*1&2*) at Kalpakkam	1000	
BSR–220 MWe/PHWRs of 700 MWe in Captive mode + BSMRs 200 MWe + BSMRs 300 MWe + SMRs 55 MWe + IPWRs-700 MWe + IPWRs-1000 MWe at various sites	43,000	Primarily sites belonging to private companies/PSUs/State PSUs
3rd Fleet of PHWRs (10 units)	*7000*	
Additional FBRs at new sites	*3000*	
Total cumulative capacity	*~ 100,000*	

it is planned to scale up the design to higher capacities to 300 MWe, 700 MWe and then 900 MWe gradually.

The technology readiness level is relatively better for capacity around 55 MWe SMR. Capability of equipment manufacturing is already established through Indian industries and efforts for proving the concepts for regulatory acceptance are likely to be minimal. The entire reactor can be barge mounted and towed and delivered to any selected site, and that is the spirit of the design. These SMRs are being targeted to captive plant for data centres, remote locations of our country lacking grid connectivity, such as Andaman and Nicobar Islands, where the cost of electricity production otherwise is exorbitantly high.

While the roadmap is ambitious however, DAE during past decades has curated its organizational structure strategically in order to achieve specific scientific and technological delivery in every possible way from its R&D units, and PSUs (Fig. 2). This has been reflected in the success story of the Stage-1 in nuclear energy scenario of the country.

At the present juncture, which would not only frame the concept but shape the path of sustainable nuclear energy scenario in the country is the commissioning of the Prototype

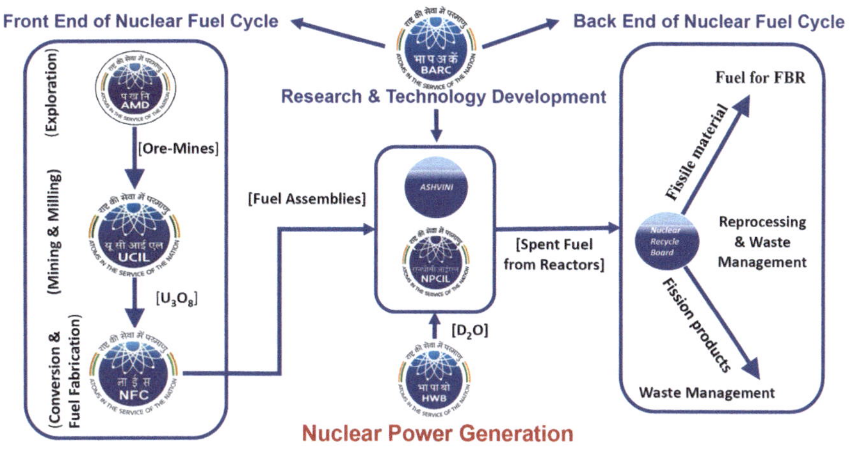

Fig. 2. Contribution of the DAE units in the success of stage-1 nuclear power by PHWR

Fast Breeder Reactor (PFBR). The department is awaiting very keenly on the combined effort of Indira Gandhi Centre for Atomic Research (IGCAR) and Bharatiya Nabhikiya Vidyut Nigam Limited (BHAVINI) to create the history by pioneering PFBR technology. Heavy Water Board (HWB) is contributing towards production of nuclear-grade sodium, as well as nuclear-grade enriched boron required for the PFBR.

As far as the PHWR is concerned, from 24 PHWRs currently operational, DAE wish to reach target of around 51 PHWRs by 2047. Over and above this number, some of PHWRs may be undertaken by public/private companies. Accordingly, all DAE units have geared up with their expansion programme to meet these challenges. With the tireless exploration efforts by Atomic Mineral Directorate for Exploration and Research (AMDR), the currently known uranium reserves are increased to 30 times higher as compared to what were known in 1970s. The unit is sparing no efforts to enhance this exploration, and targeting more and more sites where uranium could be possibly available. Presently, the Uranium Corporation of India Limited (UCIL) is providing fuels to reactors which are not in IAEA safeguards. The target is to expand this activity so that more and more PHWRs are under the domestic safeguard rather than IAEA safeguard. In regards to the fuel fabrication essentially, Nuclear Fuel Complex (NFC) has brilliantly accomplished all the targets accounted so far. As per the plan of enhancing number of PHWRs, need would arise to expand the capacity of PHWR fuel fabrication by about 6 and a half times. Consistent with this, well-defined pathway is in place both by augmenting the existing NFC facilities well as creating new facilities. NFC also plays important role in making core components such as pressure tubes, calandria tubes, and out of core components like steam generator tubes; scaling up of these activities will also be considered as necessary. Heavy water is one of the most important elements of PHWR and is produced by Heavy Water Board (HWB). India is the largest producer of heavy water in the world. Scaling up further for meeting the new requirement i.e., enhancement of about three times of its present capacity is of no technological barrier, and expansion roadmap has already been drawn.

While embarking on the Light Water Reactor (LWR) and Indian Pressurized Water Reactor (IPWR) programmes, setting up of largescale isotopic enrichment facility of uranium has also been actively considered. Recycling of spent fuel is an integral part of India's Three-Stage Power Programme. Spent fuel processing facilities for PHWR fuels are successfully established and are being operated at Kalpakkam as well as at Tarapur. A state-of-art large capacity recycling plant, named as Integrated Nuclear Recycle Plant (INRP) phase-1, is coming up at rapid pace at Tarapur. This will also facilitate reprocessing of spent fuels from 540 MWe and 700 MWe PHWRs.

For the PFBRs, MOX fuel fabrication has been carried out at BARC, Tarapur. Further a new facility for fuel fabrication is coming up at Tarapur for capacity expansion. This facility will also cater to SEU-based fuel fabrication for BSMR 200. Further, advanced plutonium fuel fabrication facility is being established at BARC, Vizag to meet the requirement of Fast Reactor Programme. Fast Reactor Fuel Cycle Facility (FRFCF) is being constructed at Kalpakkam for recycling the spent fuel from fast breeder reactors. This also includes fuel fabrication facility from recycled plutonium from fast breeder reactors. These are the glimpses of perceivable future in regard to the design, construction, commissioning of reactors, and allied activities including fuel cycle facilities, which are aiming at the 100 GWe nuclear energy goal by 2047.

In the long term, there is a third pathway for the utilization of thorium. The first is the classical three-stage program, second is through the molten salt reactor and the third route is through subcritical reactor, which is purely in research mode as on today. One way to achieve subcritical reactor is via high energy high intensity particle Accelerators Driven Subcritical (ADS) reactor, based on accelerator induced spallation reaction as against the fission reaction in the conventional nuclear reactor. Each spallation reaction, which required around 0.8–1.0 GeV energy, gives twenty-five to thirty neutrons and with such copious neutrons, ample opportunities available to transmute the fertile material into fissile, and also to incinerate the minor actinides having very long half-life. It is worth mentioning here that as the first leg in this direction, a 20 meV Low Energy High Intensity Proton Accelerator (LEHIPA) has already been commissioned in BARC (Fig. 3). The roadmap for 1 GeV accelerator constitutes four different phases such as Medium Energy High Intensity Particle Accelerator (MEHIPA)-Phase I with beam energy 40MeV, MEHIPA-Phase II up to energies of 200MeV; High Energy High Intensity Particle Accelerator (HEHIPA)-Phase III with energy level of 400MeV; and HEHIPA-Phase IV proton beam to the energy levels of 1GeV for coupling to the ADS target. This will become an important element in the future nuclear program.

Fig. 3. LEHIPA is being successfully operated at BARC

4 Challenges for Rapid Expansion: 100 GWe by 2047

Two important challenges are anticipated in achieving the targeted goal of 100 GWe by 2047. First, the capacity building would call for significant enhancement for availability of trained manpower for manufacturing, installation, commissioning, operation & maintenance, and decommissioning of nuclear plants. Efforts are being made on the outreach program, linkages with the state governments for the site and acceptability. The second important challenge is the necessity for the Indian industries to enhance their infrastructure very significantly. Indian industries need to gear up for production of high quality nuclear-grade steel, plates, piping, detectors, pumps, and other allied equipment. Current capacity of Indian industries is far below from the requirements commensurate with the target. Thus, the private industries, apart from owning the nuclear plant, also need to step up their efforts in expanding the infrastructure to cater supplies at par with the targeted capacity of 100 GWe by 2047.

5 Conclusion

The research reactor programme at BARC is a testament to India's scientific vision, technological ingenuity, and commitment to national progress. From the pioneering days of APSARA to the high-flux capabilities of Dhruva and further the advanced designs of HFRR and SMRs, BARC's R&D have driven innovation across the nuclear fuel cycle, isotope production, nuclear physics, and materials science. Through dedication, robust safety protocols, international collaborations, and a culture of excellence, DAE continues to shape India's nuclear future and contribute to global scientific advancement. As new reactors and facilities to come up, DAE is poised to sustain its leadership in nuclear research, technology, and societal impact for decades to come.

Development and Setting up of High Temperature Gas Cooled Reactor (HTGCR) and Indian Molten Salt Reactor (IMSR) Demonstration Facilities

Ved Prakash Sinha(✉)

Metallic Fuels Division Nuclear Fuels Group Bhabha Atomic Research Centre, Mumbai 400085, India
vedsinha@apsara.barc.gov.in

Abstract. The Department of Atomic Energy in its pursuit for developing advanced and green technology for energy generation and also in line with the directive of the Government of India for self-reliant (Atmanirbhar) is rigorously putting its effort towards setting up of HTGCR (High Temperature Gas Cooled Reactor) and IMSR (Indian Molten Salt Reactor) demonstration facilities. While HTGCR will primarily be responsible for producing hydrogen from a thermochemical process with direct utilization of fission heat generated in the nuclear reactor and the IMSR will be used for thorium utilization without the requirement for complex and challenging conventional pellet type nuclear fuel facilities. These reactors are also planned with the aim of achieving energy security for sustained development and growth of our country.

Keywords: Reactor · Thermochemical Process · Fast Reactor Programme · Hydrogen Utilization

1 High Temperature Gas Cooled Reactor (HTGCR)

India has set an ambitious target of becoming a Developed Nation by the year 2047 with the vision for bright and better future to its citizen. Keeping this perspective in view and also considering the ever evolving geopolitical situation, it is imperative for a country like India to safeguard its energy security for sustained and uninterrupted development. In addition, the country is majorly dependent upon massive oil imports to meet the growing energy demand and a substantial amount of forex is spent to meet the objective. The country has also demonstrated its impeccable commitment and responsibility towards environment protection, and for that it is committed to achieve net zero emissions by 2070. Therefore, keeping all views in place, the Government of India has initiated various activities and programmes like National Solar Mission, National Green Hydrogen Mission, Sustainable Alternative Towards Affordable Transportation etc. for sustained development without compromising on adverse environmental impact while ensuring energy security objective. In line to these contours, the Department of Atomic

G. B. Mohanty and V. S. Prabhudesai (Eds.): DAEC 2026, SPPHY 368, pp. 12–16, 2026.
https://doi.org/10.1007/978-981-92-2948-2_2

Energy (DAE) has also started various programmes and activities to supplement Government initiative, amongst them one of the key programme is the setting up of Small Modular Reactors (SMR's) launched under Nuclear Energy Mission. Out of the five SMR's proposed by the department, the 5 MWth High Temperature Gas Cooled Reactor (HTGCR) will be setup at Bhabha Atomic Research Centre (BARC) campus, Vizag with the objective of First of a Kind (FOAK) technology demonstration for production of hydrogen directly by utilizing fission heat generated in the nuclear reactor to a suitable thermochemical process in a nearby chemical reactor (see Fig. 1). The hydrogen produced in the reactor can be used for variety of application such as transportation, industrial (metallurgical & chemical), fertilizer etc. A successful demonstration of this reactor facility (HTGCR) will enable setting up and operation of scaled up version/higher capacity reactor units for future deployment and commercial production of hydrogen which will substantially reduce our dependence on oil import and will in turn provide massive forex savings against oil import bills.

HTGCR is a nuclear reactor facility capable of producing 5 MWth fission heat. The reactor utilizes Slightly Enriched UO_2 pellets stacked in a stainless steel clad which is surrounded in graphite sleeve. The fuel elements are arranged in specific manner and in desired configuration to get fuel assemblies. These assemblies are put inside strand pipes and are supported on reactor grid plate. Carbon dioxide (CO_2 gas) with controlled impurity content is used as process fluid for removal / extraction of fission heat from the reactor core under pressurized condition during normal reactor operation. The process fluid (CO_2 gas) while moving from bottom to top of the reactor takes away fission heat from the nuclear fuel assembly and gets collected in a hot box, which guides it to an adjacent facility (a chemical reactor) for thermochemical process. HTGCR is having twin units of chemical reactor, each with 2.5 MWth heat utilization capacity. The nuclear process heat is directly coupled with the thermochemical process for generation of hydrogen. During the chemical process CO_2 gas will cool down to a lower temperature which is further redirected and sent back to the bottom of the reactor through annular cold pipe region.

The cold CO_2 gas is used for the cooling of reactor pressure vessel and other reactor core components such as graphite blocks etc. Necessary and adequate provisions are put in place to remove fission heat from the process fluid in the event of non-availability of thermochemical reactor. Adequate safety provisions in the form of diversified shut down systems have also been put in place for safe shutdown of the reactor. A dedicated PDHR (Passive Decay Heat Removal) System and an active shut down cooling system has been incorporated in the reactor for station black out condition and removal of process heat during LOCA condition respectively. Many innovative engineering features like Customized Fuel Handling Machine for separate movement and transfer of spent fuel and fresh fuel, diversified reactivity devices, reactor and core structural material etc. will be installed and commissioned with specific objectives.

2 Indian Molten Salt Reactor (IMSR)

The luminaries and visionaries of Indian Nuclear Programme have realized in the beginning itself that Nuclear Energy has the potential to provide energy security to the country. However, they have also realized that although the country has limited Uranium reserves

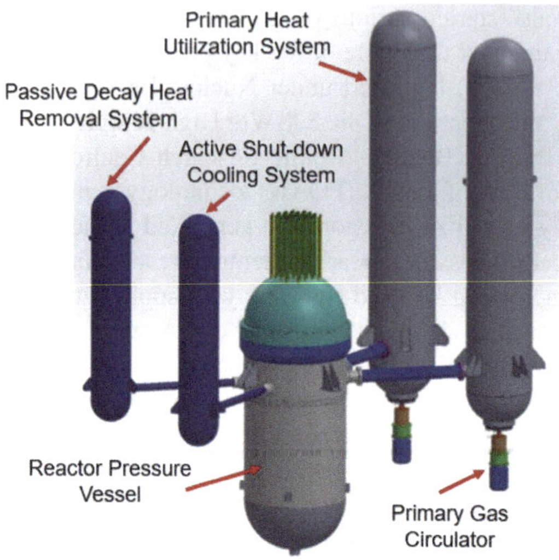

Fig. 1. Layout of 5 MWth HTGCR demonstration facility

but the Thorium is in abundance. Keeping this in view, the architect, of Indian Nuclear Programme Dr Homi J. Bhabha has proposed a three stage nuclear power programme, where in the 1st stage heavy water moderated Nuclear Reactor with natural uranium fuel will be used for energy production and during the process plutonium will be generated. The plutonium separated from the spent fuel coming out of 1st stage heavy water moderated nuclear reactor will be used in 2nd stage as driver fuel along with uranium in Fast Reactors for production of electricity and also breeding of fissile material which will enable multiplication of Fast Reactors as excess fissile material generated in the reactor will be used separately for another Fast Reactor. Meanwhile the thorium placed in the reactor blanket will be used for production of ^{233}U isotope in the 2nd stage itself. Thereafter Th-^{233}U fuel will be used in 3rd stage as driver fuel for production of electricity and making the fuel cycle self-reliant and self-sufficient (see Fig. 2).

Therefore, as discussed in the previous section and keeping the perspective of attaining country's energy security by optimum utilization of the limited available Uranium resources many programmes were initiated in the department such as irradiation of ThO_2 fuel bundle in PHWR, irradiation of thorium in Research Reactor blanket, fabrication of plate fuel with ^{233}U for KAMINI reactor etc. with a very focussed objective of attaining proficiency and self-reliance in Th-^{233}U fuel cycle. However, it was further realized that fabrication of ^{233}U fuel by classical powder metallurgy route has many challenges since the daughter product of ^{232}U are high energy gamma emitters, therefore they require engineering innovations and complexities (automation & hot cell). In view of these engineering challenges and various other reactor safety advantages, another route of utilizing thorium for energy generation and making the fuel cycle self-sustainable is explored, which is Indian Molten Salt Reactor (IMSR) (see Fig. 3). The initial studies and R&D work carried out on similar kind of system has shown potential and promise for faster movement towards energy extraction from thorium, hence a dedicated project is planned

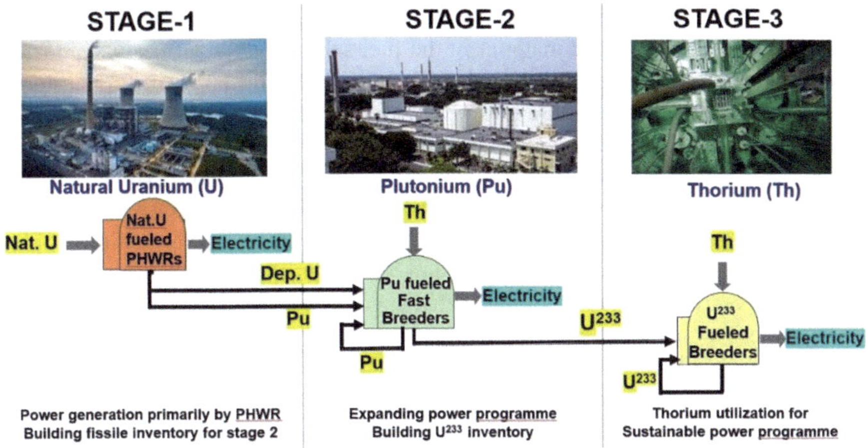

Fig. 2. Classical 3 stage nuclear programme of India

to be launched in the form of IMSR very soon. Similarly, the department is also considering another route in the form of Accelerator Driven Sub Critical System (ADSS) towards thorium utilization for energy generation. ADSS has an addition advantage of being used as facility for incineration of minor actinides other than transmutation of thorium into fissile material. In addition, ADSS has sub-critical reactor system which makes it inherently safe. Therefore, the department is in pursuit of all three parallels towards effective and efficient utilization of thorium for country's energy demand.

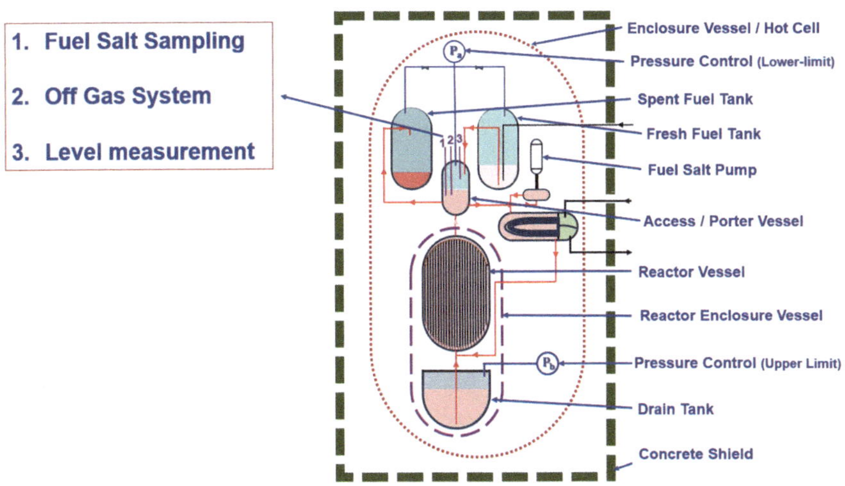

Fig. 3. Conceptual schematic of Indian Molten Salt Reactor (IMSR)

As discussed in the previous section, a demonstration reactor (IMSR) is planned to be set up at Bhabha Atomic Research Centre (BARC) campus, Vizag with molten salt as driver fuel. The fuel salt will be placed in reactor core and under the presence of thermal neutron the fissile material in the fuel salt will undergo fission to generate fission heat which will be taken out by the fuel salt itself and the fertile material (thorium) present in the salt will transmute into ^{233}U. The reactor will be having many passive safety features to make it inherently safe under all conditions including the most severe LOCA. The fission heat transported by the primary fuel salt will be extracted in a heat exchanger by using secondary salt system. The reactor will be having provision for online removal of poison and a separate arrangement will be put in place for addition of fresh reactivity for uninterrupted reactor operation. However, the promise shown by IMSR, requires successful overcoming of many challenges in the area of engineering, selection and characterization of fuel salt composition and material development. Bhabha Atomic Research Centre, Trombay is actively engaged in the development of advanced and most challenging material to meet the structural and core requirements of IMSR including primary and secondary fuel salt, engineering design of systems and the entire reactor block etc.

Acknowledgments. The author would like to place his sincere thanks and acknowledgement to all the persons, seniors and colleagues of Bhabha Atomic Research Centre, Trombay for providing all the necessary inputs in writing this article.

Bharat Small Modular Reactor (BSMR-200)

Neha Sharma(✉) and Dharmanshu Mittal

Bhabha Atomic Research Centre, Mumbai, India
{nehasharma,dharmanshu}@barc.gov.in

Abstract. An indigenous SMR, based on pressurised water reactor (PWR) technology, named Bharat Small Modular Reactor i.e. BSMR-200 having a rated power output around 220 MWe has been designed jointly by Bhabha Atomic Research Centre (BARC) and Nuclear Power Corporation of India Limited (NPCIL), based on indigenous nuclear technology envisaging delivery of equipment/components by Indian nuclear vendors. The lead nuclear power plant of BSMR, being the first of a kind with indigenous PWR, will be a demonstration unit. The paper discusses basic overview and salient design features of the BSMR-200 reactor in brief.

Keywords: Reactor · BSMR · Bharat small modular reactor

1 Introduction

Indigenous small modular reactor (SMR) programme, envisaging design and development of medium-sized nuclear power plants (NPPs), has been initiated by the Department with the broad objective of offering decarbonized captive power generation option to the domestic heavy industries sector and to repurpose existing power plant sites, deemed suitable for locating nuclear plants, vacated by retiring thermal power stations. The Small Modular Reactors (SMRs) are being developed, including the Bharat Small Modular Reactor (BSMR) as a vital component of this "Viksit Bharat" roadmap.

Bharat Small Modular Reactor i.e. BSMR-200 is an indigenous SMR having a rated power output around 220 MWe. It is based on pressurized water reactor (PWR) technology and has been designed jointly by Bhabha Atomic Research Centre (BARC) and Nuclear Power Corporation of India Limited (NPCIL). The first demonstration unit will be established at a qualified site owned by Department of Atomic Energy (DAE) i.e. Tarapur Atomic Power Station Site. The Department proposes to realize construction and operationalisation of BSMR-200 in line with the targets of the nuclear energy mission announced in the Union budget 2025–26.

BSMR-200 is the transition version based on PWR technology and it is planned to uprate the power up to 300 MWe subsequently. Design standardization will be taken up in follow-up units to ensure the cost effectiveness and project timeline optimization.

As nuclear energy can provide clean steady electric power with life cycle carbon emissions comparable to that of renewable energies, BSMR will be suitable for industries incurring large electric power consumption such as producers of steel, aluminium etc. as it will enable them to switch over to green energy options thereby averting potential penal tariffs on exports in future.

G. B. Mohanty and V. S. Prabhudesai (Eds.): DAEC 2026, SPPHY 368, pp. 17–19, 2026.
https://doi.org/10.1007/978-981-92-2948-2_3

2 General Description of BSMR-200

BSMR-200 is a PWR with rated capacity of ~ 220 MWe. Light water is used as coolant and moderator as well. BSMR-200 reactor utilizes four independent coolant circulation loops. Each loop consists of a main coolant pump and a steam generator.

The reactor core consists of 85 fuel assemblies arranged in open lattice configuration. It employs pellets of low enriched Uranium Oxide as the fuel encased in Zr1%Nb cladding. Fuel assemblies of BSMR-200 have provisions for accommodating control rods within the fuel assemblies (FAs). The control rod guide tubes are symmetrically arranged in fuel assembly according to arrangement of the control rods in the cluster assembly of control rod. Fuel assemblies are placed on the bottom grid plate in the core barrel. Core baffle is present between the core barrel shell and outer boundary of core as a thermal shield.

FAs, core barrel, core baffle and guide tube unit are placed in a Reactor Pressure Vessel (RPV) which is closed at the top by RPV head. Magnetic Jack mechanism (MJMs) i.e. drives of Reactor Control and Protection System (RCPS) are mounted on the top of RPV head. RPV head has the penetrations for connection of these drives to the movable control rod assemblies present inside guide tube unit.

3 Technological Readiness of Equipment/Components

- For reactor control and protection, the Magnetic Jack Mechanism (MJM) will be used which is pressurized and hermetically sealed electro-magnetic linear actuating mechanism. The prototype reactivity control mechanism has been developed by BARC.
- The reactor coolant pump of BSMR-200 is of shaft seal type centrifugal pump. Based on past experience with industries, the indigenous manufacturing of reactor coolant pump is feasible.
- Conventional pellets of Uranium Oxide (enrichment $< 5\%$) will be used as fuel for BSMR-200 reactor core.
- The ultraclean and high quality nuclear grade reactor pressure vessel forgings have been developed indigenously (i.e. APURVA) for Indian PWRs. The same will be utilized for BSMR-200

4 Salient Reactor Safety Aspects

4.1 Emergency Core Cooling System (ECCS)

The Emergency Core Cooling System (ECCS) is an engineered safety system designed to inject coolant into the reactor core in the event of a loss of coolant or other accident conditions that could lead to fuel overheating. The primary objective of ECCS is to maintain adequate core cooling and prevent fuel damage or core meltdown. For BSMR 200, the safety philosophy follows a combination of active and passive safety systems, ensuring redundancy and reliability even in situations where external power or operator action may not be available. The ECCS consists of the following subsystems:

4.1.1 High Pressure Injection System (HPIS)

This system injects borated water into the reactor core at high pressure during the early phase of an accident when the reactor coolant system pressure remains high. It helps maintain core cooling and provides reactivity control.

4.1.2 Passive Hydro Accumulator

These are tanks partially filled with borated water and pressurized with nitrogen. When reactor pressure drops below a predetermined level, the accumulators automatically discharge water into the reactor vessel, providing rapid and reliable core cooling without the need for active components.

4.1.3 Residual Heat Removal and Long-Term Recirculation System

After the initial injection phase, this system removes residual decay heat from the reactor core by circulating coolant through heat exchangers. It ensures long-term cooling of the core following an accident.

4.1.4 Passive Long-Term Accumulators

These systems provide additional coolant inventory to sustain core cooling over extended periods. Being passive in nature, they enhance the reliability of the ECCS by functioning without external power or operator intervention.

4.2 In-Vessel Melt Retention (IVMR)

IVMR is a strategy designed to retain the molten core material inside the reactor pressure vessel in the unlikely event of a severe core melt accident. By maintaining adequate cooling of the external vessel surface, the structural integrity of the reactor pressure vessel can be preserved, preventing vessel failure and minimizing potential challenges to the containment system. This approach significantly enhances the overall safety of the reactor.

5 Summary

Bharat Small Modular Reactor i.e. BSMR-200 is an indigenous Pressurized Water Reactor (PWR) having a rated power output around 220 MWe, being designed and developed to serve as captive power plant for domestic heavy industries. The lead unit of BSMR-200 will be established in a DAE site to demonstrate technical and commercial viability of such medium-sized indigenous PWR-based NPPs.

India's FBR Programme–Towards Realisation of Visionary's Mandate

Sriramachandra Aithal(✉), U. Partha Sarathy, and K. Natesan

Reactor Design Group, Indira Gandhi Centre for Atomic Research, Kalpakkam, India
saithal@igcar.gov.in

Abstract. To ensure the country's long term energy security, Dr Homi Jehangir Bhabha, the father of Indian Nuclear Programme, conceived a three stage nuclear power programme based on the available Uranium & Thorium resources within the country. While the first stage exploits the available limited Uranium resources, the third stage is designed to use the vastly available Thorium resources. Between these two stages, the second stage is formulated to interconnect the two stages, ensuring optimum utilisation of available Uranium resources and transition to thorium programme, smoothly. As part of this strategy, the ultimate mandate under second stage is to build fleet of fast reactors, breeding large quantity of plutonium using spent nuclear fuel from the first stage programme in the beginning years and slowly introducing thorium into these reactors to convert it into Uranium–233, which once accumulated in required quantity, will be used in third stage reactors. In line with this mandate, country's first fast breeder reactor, FBTR, a 40 MWt test reactor was commissioned at Kalpakkam in 1985. The reactor uses a UC-PuC fuel to tide over restrictions placed in importing enriched Uranium fuel as well as the limitations in Pu enrichment in Oxide based fuels. Continuing the work, Prototype Fast Breeder Reactor (PFBR), a 500 MWe, MoX fueled sodium cooled fast breeder reactor was indigenously designed, developed and installed. Though the breeding potential of oxide fuel is limited, the objective was to master the technology of large size reactor with sodium coolant, before going for metal fuel programme (with large breeding potential), which needed large scale R&D in fuel fabrication & re-processing technologies. In this quest towards reaching the ultimate objective of use of metal fuels, full scale metal fuel SA irradiation in reactor environment is identified as prerequisite for full-fledged metal fuel reactor technology. In this context, a 100 MWt test reactor, FBTR-2 has been conceptualised with sodium bonded U-23%Pu-6%Zr as test fuel and UC-PuC as driver fuel. The fuel SAs will be of same scale as that expected in 500 / 1000 MWe commercial scale reactor. After the performance of metal fuel SA is demonstrated, the core will be transitioned to full metal core to study the metal core dynamics. Besides demonstrating the metal fuel performance, the FBTR-2 will serve as replacement for FBTR, which is nearing its life, to continue the material irradiation programme, isotope production as well as demonstrate Minor Actinide (MA) burning potential. Further, pyro-reprocessing and prefabrication of reprocessed fuel is also integrated with the reactor so as to demonstrate closing of metal fuel cycle.

Keywords: India's FBR programme · Fast breeder reactors · Metal fuel

G. B. Mohanty and V. S. Prabhudesai (Eds.): DAEC 2026, SPPHY 368, pp. 20–31, 2026.
https://doi.org/10.1007/978-981-92-2948-2_4

1 Introduction

Availability of abundant and sustainable energy resources at an affordable rate is the basic requirement to achieve and sustain a developed society. Though the fossil fuels can ensure availability of cheaper energy resources, they are not sustainable due to their large contributions to global warming aspects. Further, the stock of these carbon intensive natural resources are limited. Also, India is not blessed with large resources of petroleum products. Hence, to achieve the developed status, India is required to explore other alternate resources like Nuclear and renewables. Due to the limitation in the availability in 24 × 7 mode of the renewables all-round the year, exploring the options of Nuclear is essential.

Though the known natural Uranium resources in the country is limited, India is blessed with large quantity of Thorium, the fertile material, which can be converted into a fissile material. Considering this aspect, the father of India's Nuclear Programme, Dr Homi Jehangor Bhabha conceived Three Stage Nuclear Power Programme which ensures the effective utilisation of available natural Uranium resources in first two stages whereas use of vast resources of Thorium in Third stage to ensure availability of practically everlasting energy resources for the country. While the first stage concentrates on the effective utilisation of natural Uranium in Heavy water cooled & moderated Pressurised Heavy Water Reactors, the second stage proposes to reprocess the spent fuel from the first stage PHWR reactors to extract Plutonium and feed them in a sodium cooled fast reactor to generate electricity as well as breed more plutonium from the depleted U238 fuel. This method increased the Uranium utilisation to 60–70% in comparison to ~ 0.6% expected in PHWRs. In this way, Dr Bhabha proposed that the country's energy requirements can be met while ensuring independency in energy resources. Further, the fast breeder reactors have the added advantage of higher thermodynamic efficiency as well as capability to burn minor actinides so as to bring down the radioactive waste disposal time to within the foreseeable future. With this it is possible to have a sustainable closed fuel cycle with reduced burden on deep geological storage.

To realise this mandate of Dr Bhabha, Indira Gandhi Centre for Atomic Research was established in Year 1971 (formerly, Reactor Research Centre) at Kalpakkam. The mandate of the centre was very focused to carryout Applied Research towards mastering the Fast Breeder Reactor technology and design and develop commercial scale fast breeder reactors. The details of the overall FBR programme conceived and being deployed in country is explained in following sections.

2 FBR Programme in India

India started FBR programme with the construction of FBTR. FBTR is a 40 MWt (13.5 MWe) loop type reactor and is in operation since 1985.

Subsequent to successful commissioning of FBTR, India embarked on the design & development of 500 MWe Prototype Fast Breeder Reactor (PFBR) through indigenous route. PFBR is a MoX fueled sodium cooled fast reactor. In the initial years, the conceptual design studies and optimization studies w.r.t. reactor were taken-up and completed, followed by technology development activities to ensure the manufacturability of large dimension First Of A Kind components with stringent manufacturing requirements. This manufacturing technology development exercise had helped in developing qualified vendors who could take-up manufacturing of reactor components. Presently, the construction of PFBR has been successfully completed by a specifically incorporated PSU company, BHAVINI and reactor is under commissioning.

Subsequent to PFBR, it is proposed to repeat two more units of PFBR (2 $\times$ 500 MWe) with few modifications, which are limited to incorporating feedback from PFBR design, construction, installation and commissioning. These two units will be installed at Kalpakkam complex, adjacent to PFBR.

Since, the objective of the second stage programme is to construct fleet of fast reactors, which could use the natural Uranium-238 to breed large quantity of Plutonium, to achieve effective utilisation of available natural uranium resources, higher breeding is to be targeted to multiply the reactors quickly. Since, the breeding potential of Oxide and carbide fuels are low in comparison to metal fuel, it is essential to go for the metal fueled fast reactor of commercial size like 500 MWe/1000 MWe. Presently, there are two candidate ternary metal fuels for the use, one is sodium bonded and the other one is mechanically bonded. Since, the international literature data on the metal fuel performance is limited, it is imperative to test them in a full-scale reactor environment so that the uncertainties in its performance when put in commercial size reactors can be avoided. In this context, a metal fueled test reactor of 40 MWe capacity is being conceptualized, which will also have integrated **M**etal **F**uel **C**ycle **D**emonstration **F**acility. The MFCDF will help in demonstrating the closure of fuel cycle by reprocessing the spent fuel, re-fabricating fuel pins and loading it back into reactor.

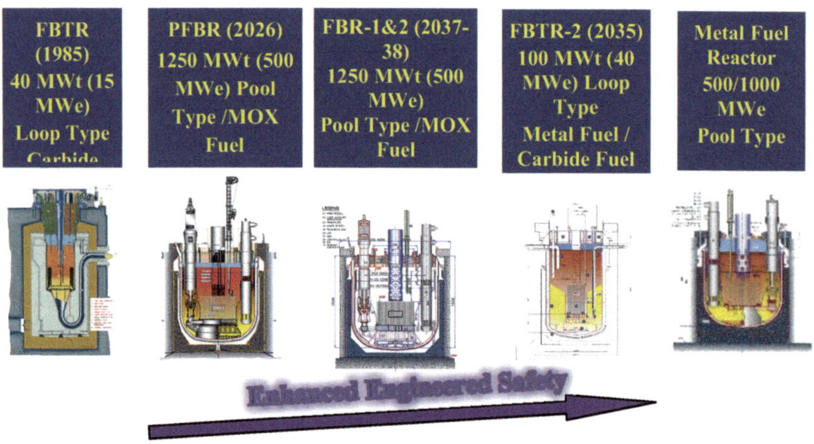

Fig. 1. Evolution of FBR programme in India

Subsequent to metal fuel technology, commercial fast reactors following FBR 1&2 will be metallic fueled reactors of 500/1000 MWe capacity, which will truly herald the maturing of 2nd stage of India's Nuclear Power Programme.

Figure 1 shows the pictorial representation of India's FBR programme. In the following paragraphs, a brief about these reactors are discussed.

3 Fast Breeder Test Reactor (FBTR)

In its quest to jump start the second stage nuclear programme, as a fore runner, India inked a cooperation agreement in early 1970s with France towards constructing & commissioning a 40 MWt sodium cooled fast breeder test reactor to provide exposure to designers, operators and sodium technologist in the field of fast reactors. The design of FBTR [1] is similar to French test reactor, Rapsodie but with the added steam cycle to produce electricity. Following were the major objectives defined for FBTR.

- To demonstrate the engineering feasibility of liquid Metal cooled fast breeder reactor (LMFBR) in India
- To provide irradiation test bed for LMFBR materials development
- To develop sodium coolant and component technology
- To develop FBR fuel reprocessing technology
- To develop a nucleus of personnel for the design, construction and operation of future fast breeder reactors

FBTR is a 40MWt, loop type, sodium cooled fast breeder test reactor. Heat generated in the reactor is removed by two primary sodium loops, and transferred to the corresponding secondary sodium loops. Each secondary sodium loop is provided with two once-through steam generator modules. Steam from the four modules is fed to a common steam–water circuit comprising a turbine-generator and a 100% dump condenser.

As envisaged in the India's Nuclear programme, important objectives of (i) Irradiation of Structural Materials, (ii) Irradiation of different fuel types, (iii) Experience with coolant systems and (iv) Safety related tests were fulfilled with the successful operation of the FBTR:

With the confidence achieved from the construction, commissioning and operation of the FBTR, India embarked on the design of much larger, commercial scale, Prototype Fast Breeder Reactor (PFBR) and the details are brought out in the next section.

4 Prototype Fast Breeder Reactor (PFBR)

PFBR is a mixed oxide (MOX) fueled, 500 MWe, sodium cooled fast breeder reactor [2]. The reactor is a pool type with two secondary loops. The primary objective of PFBR is to demonstrate techno-economic viability of fast breeder reactors on an industrial scale and indigenous design and construction capability of a FBR power plant in India. The reactor power of 500 MWe is chosen to enable adoption of a standard turbine used in fossil power stations, to have a standardized design with further reduction of capital cost & construction time in future and compatibility with regional grids. Though mixed carbide fuel of PuC-UC was used in FBTR due to nonavailability of enriched uranium for MOX fuel option, PFBR being a commercial demonstration plant, a proven fuel cycle is essential. Hence, MOX fuel is selected on account of its feasibility of developing oxide fuel with a composition as required in PFBR, proven capability of safe operation to high burnup, ease of fabrication and proven reprocessing.

Pool type concept is adopted due to its inherently high thermal inertia of the large mass of sodium in the pool which eases the removal of decay heat, use of a simple vessel with no penetrations leading to high structural integrity of the vessel and no radiation damage. Two loop design has been adopted which is economical and also meets the safety requirements. Figure 2 gives the cross-sectional details of the reactor. With the confidence from handling / operation of sodium systems in FBTR and large size engineering loops, sodium is retained as coolant for the PFBR also.

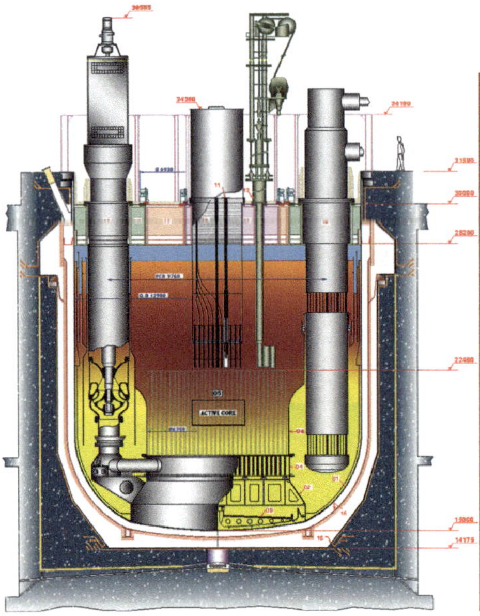

01	Main Vessel
02	Core Support Structure
03	Core Catcher
04	Grid Plate
05	Core
06	Inner Vessel
07	Roof Slab
08	Large Rotating Plug
09	Small Rotating Plug
10	Control Plug
11	CSRDM / DSRDM
12	Transfer Arm
13	Intermediate Heat Exchanger
14	Primary Sodium Pump
15	Safety Vessel
16	Reactor Vault

Fig. 2. Cross sectional details of PFBR reactor block

As adopted in FBTR, PFBR has three heat transport systems in which the primary circuit components are immersed in the sodium pool of reactor assembly. The secondary circuit also employs sodium as coolant which transfers the heat to water in the steam generator (SG). PFBR has 2 secondary loops with 4 SGs per loop. Primary circuit comprises of two primary sodium pumps and four intermediate heat exchangers. The nuclear heat generated in the core is removed by circulating sodium from the cold pool at 670 K to the hot pool at 820 K. All the reactor internals including core and primary coolant (sodium) are contained within a single vessel called main vessel. The inner vessel separates the hot and cold sodium pools. The main vessel is surrounded by safety vessel with 300 mm gap through which ISI devices will be operated to carry out inspection on the main vessel. The reactor core consists of 1758 subassemblies including 181 fuel subassemblies. The control plug, positioned just above the core, houses mainly 12 absorber rod drive mechanisms. The top of the vessel is closed with a cover known as Top Shield that includes roof slab (RS), large rotatable plug and small rotatable plug apart from forming primary sealing boundary and supporting all the components that pass through it such as primary sodium pumps, intermediate heat exchangers, control plug, transfer arm etc. The top shield also provides thermal and radiation shielding against the sodium pool in top axial direction.

5 Future FBRs

Subsequent to commissioning and demonstration of power operation, it is proposed to take-up design and construction of two more FBRs of 500 MWe capacity based on the PFBR design. In order to capitalize on the experiences gained with design, development, regulatory review, construction & commissioning of PFBR, a road map is prepared and major aspects of the road map is given below:

- Though, conceptual design of a 600 MWe FBR core with reduced sodium void coefficient was initially made, following were noted during the discussions among the stake holders:

  - The changes to static components like reactor assembly components (an increase of 1m in diameter & length) and heat exchangers (an increase of ~ 1 m in length) can be implemented without much difficulty and without losing the experiences gained.
  - However, the changes to dynamic components like pumps, shutdown & handling mechanisms calls for fresh development effort and testing before putting them in reactor.
  - Also, major changes to core & fuel design calls for detailed safety reviews, afresh.

- Hence, with an aim to minimise the design changes w.r.t. PFBR and restricting the changes to only bare essentials, which are to overcome the challenges faced during PFBR construction & commissioning, it is proposed to repeat two units of PFBR with design changes in FBR 1&2 w.r.t. PFBR, restricted to bare minimum, but found essential based on the PFBR feedback.
- Accordingly, the design of FBR 1&2 will be improved by incorporating the experiences from design, R&D, safety review, construction and commissioning of PFBR. Also, the design will be compliant to the new regulatory requirements.
- Twin reactor concept. The plant operating parameters are kept same as PFBR.
- Measures proposed to enhance safety and thereby to meet the new regulatory requirements include

  - Practical elimination of severe accident scenarios

  Stroke limiting device to eliminate unprotected transient over power accident (UTOPA)

  Hydraulically suspended absorber rods to eliminate unprotected loss of flow accident (ULOFA), if feasible to be implemented within the targeted time frame.

  - Additional secondary sodium based decay heat removal system

In the coming days, the above proposed improvements will be firmed-up based on detailed discussions with all the stake holders and further actions will be taken-up to proceed with finalization of design.

6 Metal Fueled Test Reactor: Fast Breeder Test Reactor—2

For faster growth of nuclear power generating capacity using fast reactors, large number of 500 / 1000 MWe reactors are to be built. However, with the limited Plutonium resources, the expansion of reactors in fleet mode with the oxide fuels is difficult as the reactor doubling time will be very high, of the order of 25–30 years for oxide fuels. Hence, in order to increase the generation of Plutonium and thereby install more and more numbers of FBRs, adoption of metal fuel is must as reactors with metal fuel can have shorter Reactor Doubling Time (DT) of 8 to 10 y. However, considering that there are insufficient metal fuel irradiation data in literature to go directly with full scale power reactors, it is required to test and demonstrate the metal fuel performance in a reactor environment (pin failures < 1 in 10000, inherent safety during ATWS/ULOFA, fuel utilisation of $>$ 100 GWd/t etc.) so that there are no surprises when they are adopted in large scale reactors. Though, limited scale testing in FBTR has been carried out, the length of the fuel tested is very much less (320 mm) against the power reactor requirement of 1000 mm. Also, there are inherent practical difficulties in achieving the targeted linear heat rating of 450 W/cm in FBTR core, without which fuel performance study cannot be complete. Keeping these broader requirements in mind and also considering that FBTR is ageing and has got limited useful life, it is proposed to design & construct a test reactor of ~ 100 MWt capacity as a replacement to FBTR and with following objectives:

- Testing metal fuel SA of power reactor size (1000 mm active core height) for its performance & for understanding overall core behaviour
- Core safety experiments with metal fuel to the extent possible
- Establish closed fuel cycle technologies

In addition to above listed primary objectives, following additional objectives are also identified as Secondary Objectives:

- Demonstration of large scale H_2 production (Cu-Cl cycle at 510 Deg C).
- Temperature controlled materials irradiation and testing.
- Demonstration of minor actinide incineration (MA bearing fuel testing/ Reactor physics and dynamics)
- Radio-isotope production
- Sensor calibration

6.1 Plant Parameters

Based on the objectives identified, following preliminary plant parameters (Table 1) have been worked out for the proposed test reactor:

Table 1. Preliminary plant parameters for FBTR–2

Reactor thermal power	100 MWt
Reactor inlet temperature	360 °C
Reactor outlet temperature	510 °C
IHX secondary inlet temperature	318 °C
IHX secondary outlet temperature	488 °C
Feed water inlet temperature	210 °C
Steam temperature	460 °C
Primary sodium flow rate	525 kg/s
Secondary sodium flow rate	230kg/s per loop
Feed water flow rate	21.5 kg/s per loop
Steam Pressure	125 kg/ cm^2 (a)

6.2 Core Design

Preliminary core design has been carried out with a hybrid core (Fig. 3). Since, the metal fuel of fuel column of 1m are being used first time, it was proposed to start the reactor with few metal fuel SAs driven primarily with carbide fuel. The core design was carried out with following objectives:

- Linear Heat Rating (450 W/cm)
- Burnup target: 100 GWd/t
- Safe operation & Irradiation performance–slumping, density changes, material performance etc.
- Performance of metal fuel under design basis transients.

Once the metal fuel performance is demonstrated with hybrid core, it is proposed to convert the core into full metal core to study the response of metal core under transients. However, the fuel and fuel pin design will be identical for both the hybrid and full metal core design. The salient design parameters of core are tabulated in Table 2 whereas Fig. 4 gives the schematic of proposed metal fuel subassembly.

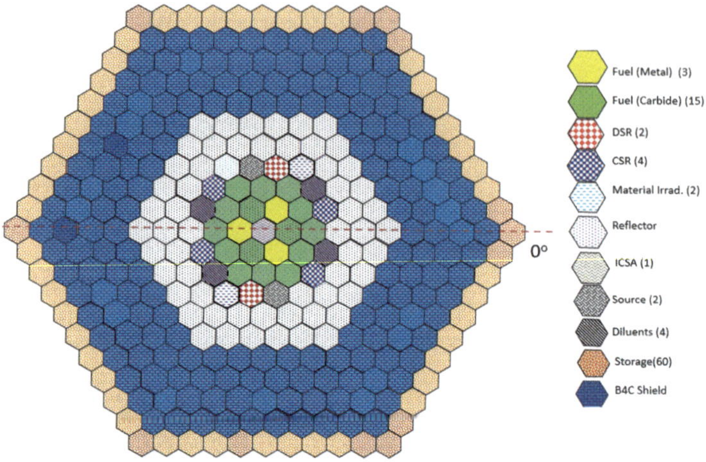

Fig. 3. Preliminary configuration of hybrid core

Table 2. Salient design parameters of core

Design parameter	Value
Reactor power	100–120 MWt
No of pins/SA	127 (Metal Fuel-MF) & 169 (Mixed Carbide-MC)
Peak LHR	470 W/cm for MC & 450 W/cm for MF
Target burn-up	100–150 GWd/t (for pin life)
Driver fuel	Mixed Carbide with 45% enrichment
Test fuel material	U-23%Pu-6%Zr
Fuel-clad gap	He bonded for MC & sodium bonded for MF
Clad & Hexcan	Clad: D9 for MC & T92 for MF; Hexcan: 9Cr-1Mo
Clad dimensions	6.6. mm for MC and 8.1 mm for MF
Fuel pellet/slug	5.4646 mm (OD-MC) 6.06 mm (OD-MF)
Fuel column height	1000 mm
Blanket column	300 mm (Bottom) /300 mm (Top)
Pin & SA length	3420 mm & 4840 mm
Max. neutron Flux	4 x 10^{15} n/cm^2-s;
DPA	100–150 (Target)

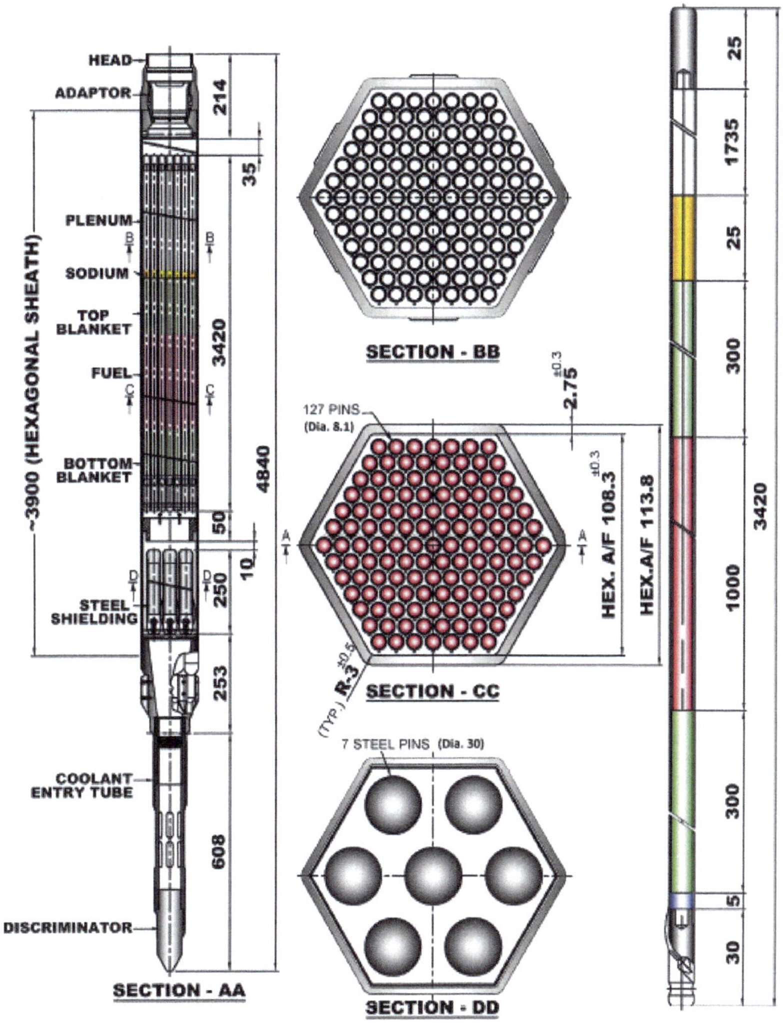

Fig. 4. Schematic of metal fuel subassembly

6.3 Conceptual Design of Reactor Assembly (RA)

A loop type configuration is proposed because of distinct advantages in terms of small size of reactor assembly (RA can be made transportable, in few axial segments), reduced complexity of top shield, reduced construction time, ease of operation, and improved maintainability, while adequately meeting the safety requirements of a low-power test reactor such as FBTR-2. In addition, the loop concept aligns well with prevailing international practices and benefits from substantial domestic operational experience gained from FBTR. However, along with the simplification of RA, loop concept also demands additional elaborate primary circuit outside the RA, housed inside shielded cells. For the loop type concept, Reactor Assembly has been conceptualized with two inlet and two outlet pipes. Nevertheless, the pool type concept is also being studied for its merits over the loop type.

Overall reactor assembly cross section for FBTR-2 is shown in Fig. 5. Important design features considered in reactor assembly are multiple supports paths for core support structure, Nozzle penetration in cover gas space below the top shield to reduce complexity over top shield as well as reduce the operating temperature of nozzle, core catcher to cater to the full core meltdown etc. Also, feasibility of removal of main vessel cooling system is also being studied, since maximum design temperature is 510 °C.

With the preliminary sizing, diameter of main vessel is worked out to be 7480 mm with MV cooling whereas 7110 mm without MV cooling. The height of the RA is worked out to be 13224 mm. Considering active core dimension, control plug size (dia.1150 mm), a handling radius of 1220 mm, two rotatable plugs concept, the choice of single straight pull machine with charging/discharging flask for handling of core sub-assemblies and similar sealing concept as that used in PFBR, the SRP diameter is obtained to be 3525 mm whereas LRP diameter is obtained to be 5700 mm.

The heat transport system for the FBTR-2 has been conceptualized and there are two primary loops and two secondary loops (one per each of the primary loop). The expected maximum core outlet temperature is 510 Deg C whereas secondary sodium temperature at the inlet to steam generator will be 488 Deg C. The steam will be produced with the parameters of 460 Deg C and 125 bar pressure.

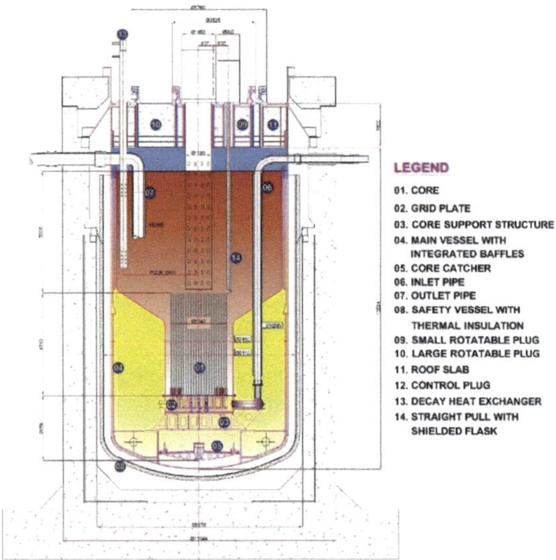

Fig. 5. Cross sectional details of FBTR-2

7 Summary

The article briefly presents the background for the India's three stage nuclear power programme and the necessity of fast breeder reactors to ensure energy security to the country. The evolution of India's FBR programme is highlighted along with the brief description of different fast reactors which are in operation, commissioning and planning. The importance of metal fuel programme is highlighted in the context of achieving higher breeding ratio, leading to quicker multiplication of fast reactors.

8 Competing Interests

The authors have no competing interests to declare that are relevant to the content of this article.

Acknowledgments. Authors are thankful to Chairman, AEC and Director, IGCAR for nominating the first Author to attend the 2026 version of DAE Conclave and present on the 2nd stage of India's Nuclear Power Programme. Authors are also thankful to the organisers of the DAE Conclave—2026 for providing an opportunity to present the details and publish the presentation details.

References

1. G. Srinivasan, K.V., Suresh Kumar, B., Rajendran, P.V.: Ramalingam the fast breeder test reactor—Design and operating experiences. Nuclear Eng. Des. **236**, 796–811 (2006)
2. Chetal, S.C., Balasubramaniyan, V., Chellapandi, P., Mohanakrishnan, P., Puthiyavinayagam, P., Pillai, C.P., Raghupathy, S., Shanmugham, T.K., Sivathanu Pillai, C.: The design of the prototype fast breeder reactor. Nuclear Eng. Des. **236**, 852–860 (2006)

Setting-Up of an Isotope Production Reactor

Kapil Bodkha(✉), Monesh Chaturvedi, Jit Pal Singh, and J. Chattopadhyay

Bhabha Atomic Research Centre, Mumbai, India
kbodkha@barc.gov.in

Abstract. Radioisotopes have widespread applications across several sectors, with one of the most important uses in healthcare, where they are extensively employed for diagnosis and therapy of cancer. With the growing demand for medical radioisotopes and ageing production facilities, it is essential to establish a dedicated facility capable of ensuring sustained and reliable supply. Such a facility would meet domestic requirements for years to come while also creating opportunities for export. To address this need, a dedicated high-flux Isotope Production Reactor (IPR) is planned for large-scale production of radioisotopes primarily for healthcare applications. The reactor is designed to deliver high thermal neutron flux, enabling efficient and continuous isotope production. The IPR is expected to strengthen the domestic supply chain of medical radioisotopes, reduce dependence on imports, and enable production of specialized radioisotopes.

Keywords: Reactor · Radioisotope · Healthcare

1 Introduction

The growing role of radioisotopes across healthcare, industry, agriculture, and research has led to a significant increase in global demand, particularly in healthcare sector involving use of nuclear medicine for diagnosis and therapy, and in general, for equipment sterilization. Modern techniques in healthcare such as nuclear imaging, curative and palliative radiotherapy, and tracer studies rely heavily on a steady and reliable supply of medical radioisotopes. In the context of India, demands of radio-isotopes in healthcare is continuously increasing owing to increase in cancer loads. In view of huge population of the country, even a small percentage increase in patients results in huge demand of supporting infrastructure and facilities. To handle the rising patient loads efficiently, there is a need for the early and precise diagnosis of the disease. At the same time affordable diagnosis and treatment is topmost priority as it can cover a large spectrum of patient base.

Besides sourcing infrastructure (like radioisotope production facilities), a significant development of nuclear medicine infrastructure is also essential. As of 2023, India operated 442 nuclear medicine centers, 359 PET-CT scanners and 24 medical cyclotrons while 150 high dose radionuclide therapy facilities were functional [1]. To meet the growing clinical demand and achieve equitable access, substantial expansion of imaging and radionuclide therapy facilities is required over the next decade.

G. B. Mohanty and V. S. Prabhudesai (Eds.): DAEC 2026, SPPHY 368, pp. 32–35, 2026.
https://doi.org/10.1007/978-981-92-2948-2_5

Radioisotopes are produced through multiple routes, including research reactors, power reactors, accelerators and chemical separation of reprocessed materials. Key medical isotopes such as I-131, Mo-99, Lu-177, Ir-192 and Co-60 are predominantly produced in a reactor, while accelerators support production of radioisotopes like F-18, Cu-64, Tl-201 and Ga-67. Useful radioisotopes like Cs-137, Ru-106 and Sr-90 are obtained by chemical separation from waste generated during reprocessing.

Presently, a total of 228 research reactors are operating worldwide in 54 countries [2]. However, fewer than 10 reactors supply most of the medical radioisotopes to other nations. The global supply chain remains vulnerable, as these reactors are ageing with many expected to retire by 2030. This may be leading to a demand-supply mismatch. In this context, it is planned to set-up a new high flux reactor dedicatedly for the production of radioisotopes. The IPR is expected to significantly strengthen India's nuclear medicine capacity, support self-reliance in radioisotope production, and contribute to meeting both domestic and global healthcare needs in the coming decades.

2 General Description of IPR

IPR is a 40 MW open pool type reactor with a high thermal neutron flux of approximately 1.0×10^{15} n/cm^2/s. It employs LEU in the form of U_3Si_2 in aluminium matrix as fuel, and light water as coolant as well as moderator. The reactor core consists of plate type fuel assemblies (22 standard fuel assemblies and 6 control fuel assemblies). The reactor core and associated structures like plenum structure, grid plate, grid plate support, reflector vessel and chimney are located inside reactor pool (RP) filled with de-mineralized light water.

The reactor core chimney is open at the top to the RP and surrounded by an annular reflector tank filled with heavy water as reflector. The reflector tank is supported on a cylindrical inlet plenum which is supported on bottom of the RP. Inside the inlet plenum, a cylindrical mechanical filter is provided which act as filter cum diffuser for the coolant entering the core. The inlet plenum forms a close pressure boundary to the primary heat transport system (PHTS) of the reactor and serves as a coolant header. The top of the inlet plenum is connected (not hydraulically) with irradiation plenum of the reflector vessel by bolted joint. The light water coolant enters the inlet plenum after it passes through a mechanical filter and then enters the fuel assemblies and exits from the chimney through two outlet nozzles.

The reflector region provides large volume for irradiation which forms majority of the irradiation locations. Irradiation locations are also provided in the core region for production of high specific activity radioisotopes. The target materials for irradiation are kept in Aluminium capsules inside tray rods and the tray rods are placed at different irradiation locations for production of the isotopes. The isotopes that are planned in IPR are provided in Table 1. IPR also employs a tray rod facility which essentially is a hot cell and has manipulators for removing irradiated capsules from tray rods and collecting them in a shielded cask. The cask is then transferred to the isotope processing lab for further processing.

Table 1. Radioisotope production planned in IPR

Isotopes	Annual capacity (Ci)	Sp. activity (avg.) (Ci/g)
Ir-192	4,80,000	1000
Co-60 (HSA)	35,000	300
Mo-99/Tc-99m	60,000	4
Lu-177	48,000	35,000
Lu-177 (NCA)	8000	80,000
I-131	36,000	60,000
Sm-153	1600	5000
Ho-166	1600	2500
W-188/Re-188	400	20
Y-90	150	10
Co-60 (Ind)	75,000	30

3 Reactor Process Systems

IPR employs different process systems to remove heat generated in various regions (core, reactor pool, reflector region etc.) and dissipate the same to the ultimate heat sink via induced draft cooling tower (IDCT).

The core heat is removed by circulating light water supplied by two loops of primary heat transport system (PHTS). In order to ensure safe and stable operation of the reactor, it is important to avoid boiling in the core. To fulfil this requirement, a sufficient margin is kept from onset of nucleate boiling (ONB) by maintaining sufficient velocity in the flow channels. Ensuring ONB margins during normal operations also aligns other margins (like CHF) within limiting values. The PHTS, with two flywheel mounted pumps in each of the two loops maintains the flow through the core. Each loop of PHT also employs two plate type heat exchangers to transfer the heat generated in the core to the service water employed on the secondary side. Under shutdown conditions (including those in LOCA), the core cooling is maintained with the flow imparted through class-II power backed shutdown cooling pumps (SDCPs). There are two trains of the shutdown cooling system, each employing two SDCPs (including one for redundancy) and a shutdown heat exchanger. In the event of prolonged shutdown or when SDCPs are also not available in SBO due to exhaustion of battery, the provision of natural cooling of core is made by providing two normally closed natural circulation valves, one each in each of the coolant inlet line connecting to the inlet plenum.

Heat generated/deposited in the RP along with the decay heat of the spent fuels is removed with the help of pool water cooling system. For the planned shutdown or under accidental conditions the pool water is cooled by long term pool water cooling systems (LT-PWCS) which employs one class-III + DEC-DG backed pump in each of the two independent trains. Reflector cooling system comprising of canned motor pumps and shell & tube heat exchangers, cools the heavy water inventory and maintains its temperature within limits. Heat from all the above mentioned systems is transferred to service water systems through respective heat exchangers for dissipation to the ultimate heat sink. PHTS, PWCS and RCS are provided with delay tank for decaying the N-16 activity.

4 Reactor Safety Features

The reactor incorporates a range of inherent and engineered safety features to ensure safe operation under normal and off-normal conditions. It is designed as a low-enthalpy system with natural circulation cooling, enabling reliable decay heat removal during loss of power supply without dependence on active systems.

The IPR is provided with two independent shutdown systems. The primary shutdown system (PSS) consists of six control cum shut off rods (CSRs). Emergency shutdown system (ESS) consists of four emergency shut-off rods along with reflector dumping system. The CSRs are also used for the control of the reactor power. Apart from that, a hot water layer is maintained at the top of the reactor pool to establish thermal stratification, which limits large-scale mixing of pool water and thereby restricts the upward transport of radioactive water, reducing the likelihood of radioactivity reaching the pool surface. Reactor safety is further enhanced by a strong negative coefficient of reactivity, providing inherent power self-regulation. Additional safety provisions include overhead water tanks for gravity-driven emergency cooling, redundant main and supplementary control rooms for assured monitoring and control, and underwater handling of spent fuel and irradiation tray rods to ensure effective radiation shielding.

5 Summary

The IPR is conceived as a dedicated, high-flux reactor designed to address the growing demand for medical radioisotopes and to strengthen India's nuclear medicine ecosystem. The IPR is expected to significantly reduce dependence on imported radioisotopes, enable the production of specific-purpose isotopes, and create opportunities for export. Overall, the project represents a strategic investment in healthcare infrastructure, supporting affordable cancer diagnosis and treatment.

References

1. Singh, B., Sahani, G., Prasad, V., Watts, A., Kumar, R.: India's growing nuclear medicine infrastructure and emergence of radiotheranostics in cancer care: associated challenges and the opportunities to collaborate. Indian J. Nucl. Med. **38**(3), 201–207 (2023)
2. IAEA Research Reactor Database: https://nucleus.iaea.org/rrdb/home. Last accessed on 2026/02/11

Quantum Technology

Atomic Vapor Based Quantum Sensors for Electromagnetic Fields Spanning DC to THz Frequencies

Nandini Mondal[1,2], Soumya Ranjan Mishra[1,2], Md. Tousif Alam[1,2], Surya Narayan Sahoo[1,2], and Ashok Kumar Mohapatra[1,2(✉)]

[1] School of Physical Sciences, National Institute of Science Education and Research Bhubaneswar, Jatni 752050, India

[2] Homi Bhabha National Institute, Training School Complex, Anushaktinagar, Mumbai 400094, India

a.mohapatra@niser.ac.in

Abstract. Atomic vapor based quantum sensors have emerged as a powerful platform for the precise and broadband measurement of electromagnetic (EM) fields. The strong interaction between external electric fields and highly excited Rydberg-Rydberg transitions enables calibration-free detection of electric fields with high sensitivity. In particular, Rydberg atom–based techniques combined with all optical excitation to the Rydberg state such as electromagnetically induced transparency (EIT), allow accurate field measurements over an exceptionally wide frequency range spanning from DC to the terahertz (THz) regime. However, in the low-frequency range (DC to MHz), the sensitivity of Rydberg atom–based sensors is limited and does not surpass that of classical sensors. In contrast, atomic magnetometers are better suited for achieving higher sensitivity in the radio-frequency regime. This work reviews the underlying principles, experimental implementations, and recent advances in Rydberg atom–based sensors, which operate efficiently in the GHz to THz regime, as well as atomic magnetometers based on the magnetoelectric effect in atomic vapor for sensing in the DC to MHz regime.

1 Introduction

Quantum sensors based on atomic and optical physics are opening new avenues in technology and precision metrology, with potential implications for fundamental physics [1]. Quantum-enhanced electric field sensors, in particular, hold significant promise for applications in communications and remote sensing [2]. Electric field sensing based on electromagnetically induced transparency (EIT) in microwave-dressed Rydberg atoms [3] has demonstrated excellent sensitivity in the GHz and THz frequency bands [4,5]. However, in the MHz regime, Rydberg atomic sensors typically operate in an off-resonant regime, leading to limited sensitivity [2,6]. In contrast, atomic magnetometers based on magnetic

G. B. Mohanty and V. S. Prabhudesai (Eds.): DAEC 2026, SPPHY 368, pp. 39–48, 2026.
https://doi.org/10.1007/978-981-92-2948-2_6

dipole interactions, provide an alternative pathway for low-frequency detection. These interactions can be exploited to sense the magnetic field component of radio-frequency waves, making them well-suited for applications in the DC to MHz regime.

Over the past decade, atomic magnetometers have achieved remarkable sensitivities at the femtotesla level within compact form factors, making them highly attractive for a wide range of industrial and biomedical applications [7–13]. Optically pumped atomic magnetometers (OPMs) exploit the interaction between light and alkali atomic vapors, wherein an external magnetic field induces a rotation in the polarization of a probe beam [14]. This polarization rotation measured using balanced detection method serves as a highly sensitive measure of the applied magnetic field. OPMs based on this principle have demonstrated exceptional sensitivities reaching the sub-femtotesla regime [15]. Notably, such performance has been achieved even in unshielded environments, in both DC and radio-frequency (RF) domains [16–18], thereby enabling a wide range of in-situ and real-world sensing applications.

Atomic magnetometers utilizing the nonlinear magnetoelectric effect (NME) [19] offer sensitivities ranging from the picotesla to femtotesla level over a broad radio-frequency bandwidth, along with a significantly enhanced dynamic range. The NME is a wave-mixing process involving an optical electric field oscillating at frequency ω_p and a magnetic field oscillating at radio frequency ω_{rf}. The corresponding second- and third-order nonlinear polarization components can be expressed as $\chi_{ijk}^{eem} E_j B_k$ and $\chi_{ijkl}^{eemm} E_j B_k B_l$, respectively. These processes give rise to newly generated optical fields at frequencies $\omega_p + \omega_{rf}$ and $\omega_p + 2\omega_{rf}$. Detection of these generated optical sidebands using heterodyne techniques provides a clear signature of the presence of the RF magnetic field. In single-beam magnetometry based on parametric processes such as the nonlinear magneto-electric effect [19] is particularly advantageous, offering high sensitivity due to the coherent detection scheme, along with a large dynamic range and arbitrary frequency resolution of the RF field.

2 Rydberg Atom Sensor for EM Fields

The operating principle of the Rydberg atom–based electromagnetic field sensor is illustrated in Fig. 1. The relevant energy-level scheme of rubidium is shown in Fig. 1(a). A probe laser at 780 nm drives the $5S_{1/2} \rightarrow 5P_{3/2}$ transition, while a coupling laser at 480 nm excites atoms to the Rydberg state via the $5P_{3/2} \rightarrow 62S_{1/2}$ transition. Under simultaneous application of the probe and coupling fields, electromagnetically induced transparency (EIT) renders the medium transparent to the probe beam. This effect enables coherent, all-optical detection of Rydberg excitations in thermal vapor [3]. A microwave (MW) field couples the $62S_{1/2} \rightarrow 62P_{3/2}$ transition, with a resonant frequency near 15 GHz. In the presence of the MW field, the Rydberg-Rydberg transition undergo Autler–Townes splitting, characterized by the Rabi frequency $\Omega_{\text{mw}} = \frac{2\mu E}{\hbar}$, where μ is the dipole matrix element of the Rydberg transition, E is the MW electric-field amplitude, and $\hbar$ is the reduced Planck constant. The resulting modification of the EIT spectrum is shown in Fig. 1(b). In the absence of the MW field, a single transparency peak is observed. When the MW field is applied, a clear Autler–Townes doublet emerges, with the splitting directly proportional to the MW electric-field amplitude, thereby enabling absolute, calibration-free field measurement. Notably, the resonant frequencies of Rydberg–Rydberg transitions span the GHz to THz regime, enabling broadband detection of electromagnetic fields. This wide tunability is achieved by selecting appropriate Rydberg states via suitable excitation laser frequencies.

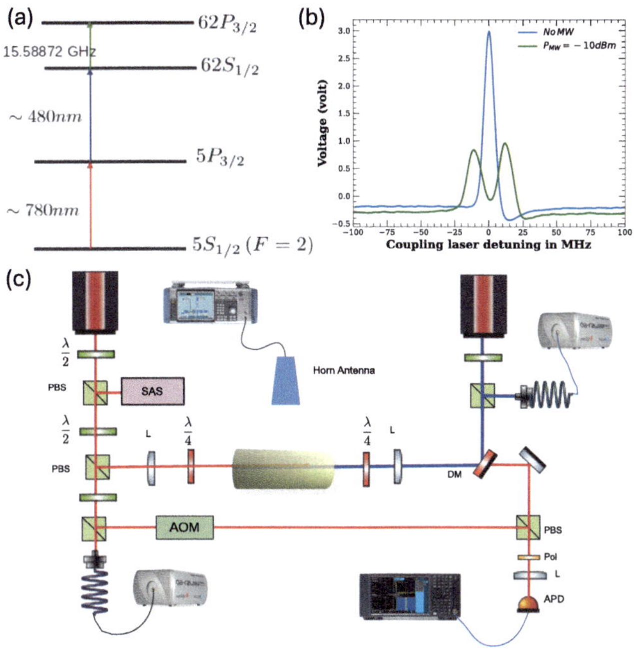

Fig. 1. (a) Energy-level diagram of rubidium showing the excitation scheme to the Rydberg state. (b) Rydberg EIT spectra in the absence (blue) and presence (green) of the microwave field. The microwave field induces a clear Autler–Townes splitting of the EIT resonance. (c) Schematic of the experimental setup.

The schematic of the experimental setup is shown in Fig. 1(c). A 5 cm-long rubidium vapor cell is heated to ∼100° C. The probe (780 nm) and coupling (480 nm) beams, derived from commercial diode laser systems, are aligned in a counter-propagating configuration and focused into the vapor cell using 50 cm focal length lenses. Quarter-wave plates are employed to generate mutually orthogonal circular polarizations. The probe laser is frequency-locked to the $5S_{1/2} \longrightarrow 5P_{3/2}$ (D_2) transition of rubidium via saturated absorption spectroscopy. To enhance the signal-to-noise ratio, the coupling beam is amplitude-

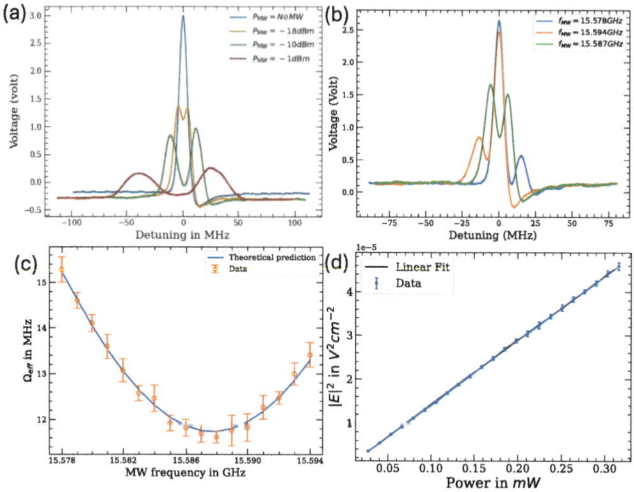

Fig. 2. (a) Autler–Townes splitting of the Rydberg EIT signal for different on-resonance microwave field amplitudes. The solid curves in magenta, green, and red correspond to microwave powers of -18, -10, and -1 dBm, respectively, while the blue curve shows the EIT signal in the absence of the microwave field. (b) Autler–Townes splitting for off-resonant microwave fields. The blue, green, and magenta curves correspond to microwave frequencies of 15.578 GHz, 15.594 GHz, and 15.587 GHz, respectively. (c) Effective Rabi frequency as a function of microwave frequency. Open circles represent measured values extracted from the Autler–Townes splitting, with error bars indicating the standard deviation from repeated measurements; the solid line is a theoretical fit. (d) Squared electric-field amplitude at the interaction region as a function of source power. The solid line represents a linear fit.

modulated at 5 kHz using a mechanical chopper, and the transmitted probe signal is detected using a lock-in amplifier. The coupling laser frequency is scanned to obtain the Rydberg EIT spectrum. The microwave field is generated using a vector signal generator (SMM100A, Rohde and Schwarz) and applied to the vapor cell via a horn antenna kept at a distance of 30 cm from the vapor cell. Both laser frequencies are monitored using wavelength meters.

The experimental results are presented in Fig. 2. The Autler–Townes splitting increases with microwave power, as shown in Fig. 2(a). For resonant microwave excitation of the Rydberg–Rydberg transition, the splitting is symmetric about the line center. In contrast, when the microwave field is detuned from resonance, the splitting becomes asymmetric, as shown in Fig. 2(b). In the detuned case, the splitting is governed by the effective Rabi frequency, $\Omega_{\text{eff}} = \sqrt{\Omega_{\text{mw}}^2 + \Delta_\mu^2}$ where Δ_μ is the microwave detuning from the Rydberg–Rydberg transition. The effective Rabi frequency extracted from the measured splitting is plotted as a function of microwave frequency in Fig. 2(c), and fitted to the above expression to determine Ω_{mw}. The electric-field amplitude is then obtained from $\Omega_{\text{mw}} = \frac{2\mu E}{\hbar}$. The resonant frequency of the Rydberg-Rydberg transition is also determined to be

15.58872 GHz. Notably, this approach enables absolute electric-field measurement without external calibration, as it relies solely on the measured splitting and fundamental atomic constants. The squared electric-field amplitude at the interaction region as a function of source power is shown in Fig. 2(d), exhibiting a linear dependence. Losses due to cable transmission and free-space propagation do not affect the measurement, since the technique directly probes the local field amplitude. Owing to its dependence only on atomic parameters, the sensor can serve as a traceable standard for calibration of electromagnetic field sensors.

3 RF Magnetometer Based on Nonlinear Magnetoelectric Effect in Atomic Vapor

The nonlinear magnetoelectric effect (NME) arises from the parametric interaction of optical and radio-frequency (rf) fields via multi-wave mixing processes, resulting in the generation of new optical fields. Suppose a dc magnetic field applied is applied in counter-propagating direction of the pump light. Taking the pump light propagation direction as quantization axis, the Zeeman energy levels in the ground state ($F = 1$) is splitted as shown in Fig. 3(a) and (b). The excited state is taken $F' = 0$. A σ^+-polarized pump field of frequency ω_p couples the ground state $m_F = -1$ to the excited state $m'_F = 0$, leading to optical pumping of population into the $m_F = 0$ and $m_F = +1$ ground states. The rf magnetic field enables coherent coupling between ground-state Zeeman sublevels. In the three-wave mixing process, an atom initially in the $m_F = 0$ state emit a σ^+ rf photon to the $m_F = -1$ state, followed by absorption of a σ^+ pump photon to the excited state $m'_F = 0$, and subsequently emits a π-polarized optical photon returning to $m_F = 0$. This process gives rise to a second-order nonlinear polarization, $P_\pi^{(2)} = \chi^{eem}_{\pi\sigma^+\sigma^+} E_{\sigma^+} B_{\sigma^+}$ and generates an optical field at frequency $\omega_{g1} = \omega - \omega_{rf}$. In the four-wave mixing process, an atom initially in the $m_F = +1$ state emits two σ^+ rf photons to $m_F = -1$ state and absorb one σ^+ pump photon to reach the excited state $m'_F = 0$, and then emits a σ^--polarized optical photon returning to $m_F = +1$. This process is described by a third-order nonlinear polarization, $P_{\sigma^-}^{(3)} = \chi^{eemm}_{\sigma^-\sigma^+\sigma^+\sigma^-} E_{\sigma^+} B^2_{\sigma^+}$ and generates an optical field at frequency $\omega_{g2} = \omega - 2\omega_{rf}$. Neglecting the wave vector of rf-photon, the phase-matching conditions determine the propagation direction of the generated fields along the pump light.

A schematic of the experimental setup is shown in Fig. 3. The pump beam, derived from an external-cavity diode laser operating at 780 nm, is split into two parts. One part is directed to a 5-cm-long Rb vapor cell, while the other is frequency-shifted by 40 MHz using acousto-optic modulators to serve as a local oscillator (LO). The input pump polarization is controlled using a quarter-wave ($\lambda/4$) or half-wave ($\lambda/2$) plate to produce circular or linear polarization, respectively, before the cell, and converted back to linear (or orthogonal linear) polarization after the cell. Since the generated fields propagate collinearly with the pump beam, all output fields are combined with the LO using a polarizing beam splitter. The combined signal is passed through a polarizer and detected

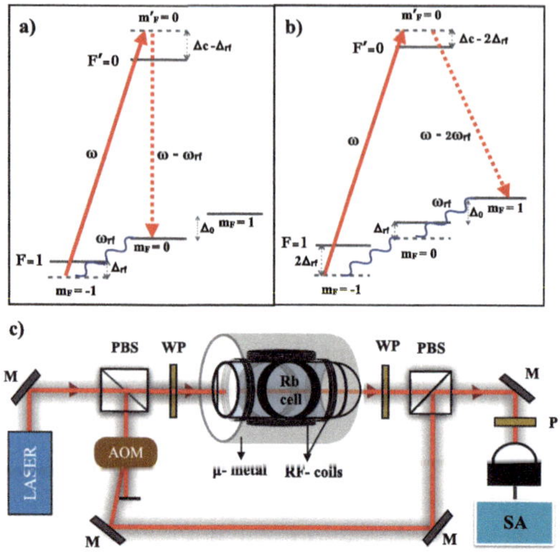

Fig. 3. Schematic of the energy level diagram for D_2 line, ^{87}Rb $F = 1 \rightarrow F = 0$ transition. a) Depiction of the three-wave mixing process and b) four-wave mixing process in the system. The input fields are the pump (ω) and an rf field (ω_{rf}) leading to the optical generation at frequencies, ω-ω_{rf} via three-wave mixing and ω-2ω_{rf} via four-wave mixing process. Δ_c (Δ_{rf}) is the detuning of the input optical (rf) field from the corresponding atomic transition. c) Schematic of the experimental setup for the observation of the mixing process. The details can be found in reference [19]

using a fast photodetector, and the resulting spectrum is analyzed using a spectrum analyzer. Three orthogonal Helmholtz coil pairs surround the vapor cell along the x, y, and z directions. A DC current is passed through the pair of Helmholtz coils aligned along the z-direction to produce a static magnetic field in the vapor cell. Because the coil dimensions are finite relative to the cell length, the magnetic field is inhomogeneous along the interaction region. This inhomogeneity allows both second and third-order processes enabling the generation of light at both ω_{g1} and ω_{g2}. An oscillating magnetic field of amplitude 0.14 G is applied along the x-axis. The vapor cell is shielded from stray magnetic fields using three layers of μ-metal and is heated to 80°C, corresponding to an atomic density of 1.8×10^{12} cm^{-3}. The pump power at the entrance of the cell is 10 μW, with a beam waist of ~1.5 mm. The mixing processes are most efficient when the rf drive frequency ω_{rf} matches the Zeeman splitting Δ_0. Maximum signal is observed for a laser detuning of 300 MHz to the red of the ^{87}Rb $F = 1 \rightarrow F = 0$ transition. Experimental spectra for σ^-, σ^+, and π input polarizations are shown in Fig. 4(a)–(c), respectively. The central peak at 40 MHz arises from interference between the pump and LO fields. For σ^+ (σ^-) polarization, the generated sidebands appear at lower (higher) frequencies relative to the pump, resulting in peaks on the left (right) side of the central peak. For π-polarized input, symmetric sidebands are observed on both sides. In the absence of the rf magnetic field, no sidebands are present, as shown in Fig. 4(d). The appearance of these

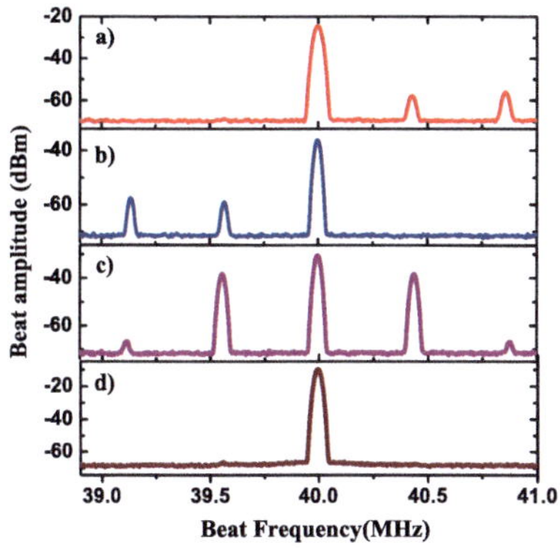

Fig. 4. Experimental data of the generated beat amplitudes for input a) σ^+ light, b) σ^- light, c) π polarised light in the presence of rf magnetic field and d) without rf magnetic field.

sidebands serves as a clear signature of the rf magnetic field, forming the basis of rf magnetometry.

The sensitivity is evaluated as $\delta B_{rf} = \frac{\delta S}{\left(\frac{dS}{dB_{rf}}\right)}$ where $\frac{dS}{dB_{rf}}$ is obtained from a polynomial fit to the detector signal as a function of B_{rf}. The uncertainty in the normalized spectral amplitude, δS, is scaled by $\sqrt{1/T}$, where T is the acquisition time. For the three-wave mixing (second-order) process, the detector signal exhibits a linear dependence on B_{rf}, rendering $\frac{dS}{dB_{rf}}$ independent of B_{rf}. In contrast, for the four-wave mixing (third-order) process, the detector signal depends quadratically on B_{rf}, and hence the sensitivity becomes B_{rf}-dependent. The measured sensitivity for the second-order remains independent of B_{rf} and characterised over a broad frequency range from 10 kHz to 3 MHz [19]. The peak sensitivity is 70 fT/$\sqrt{Hz}$ at 10 kHz and pT/$\sqrt{Hz}$ at 1 MHz [19]. As depicted in Fig. 5, the detector signal shows an overall quadratic dependance on B_{rf} for third-order process, with deviations attributed, the detector signal shows an overall quadratic dependence on B_{rf}, with deviations attributed to higher-order processes involving multiple rf-induced transitions between the same Zeeman sublevels. The sensitivity is determined from the local slope at different values of B_{rf} along with the corresponding signal uncertainty. The best experimentally obtained sensitivity is ≈ 6.25 pT/$\sqrt{\text{Hz}}$. At low B_{rf}, the sensitivity is limited by photon shot noise of the generated optical field, while at higher amplitudes it is constrained by fluctuations in the applied rf magnetic field itself.

The sensor exhibits arbitrary frequency resolution for rf magnetic-field detection due to the parametric nature of the underlying mixing process. In addition, a bandwidth of ~200 kHz is observed. These features enable the use of the system as an rf magnetic-field spectrum analyzer, as shown in Fig. 6. To demonstrate this capability, the rf magnetic field is modulated with a square wave having a fundamental frequency of 10 kHz. As expected, the even harmonics are sup-

Fig. 5. The normalized spectral amplitude (solid blue line) as function of the RF (100 kHz) magnetic field amplitude (estimated using [20]). The polynomial function of even orders ($a_2x^2 + a_4x^4 + a_6x^6 + a_8x^8$) fits the data with the coefficients $a_2 = 0.215, a_4 = -0.014, a_6 = 0.000832$ and $a_8 = -0.0000233$. The error is multiplied by a factor of 25 for representation. The measured sensitivity is given by the solid green line. The details of the sensitivity measurement for 2nd order case can be found in reference [19].

pressed, while the odd harmonics agree well with the theoretical amplitudes of an ideal square wave.

In summary, we present atomic vapor based quantum sensors which can detect electromagnetic fields spanning the range from DC to THz. For high frequency (GHz to THz) Rydberg atom based sensor using the dipole coupling of electric field amplitude is sensitive. The Rydberg atom based sensor doesn't require calibration and gives the absolute measurement of the electric field amplitude at the interaction region. hence it can be used as a traceble standard. However, the Rydberg atom sensor is not sensitive at low to mid frequency range. We have demonstrated NME based atomic magnetometer with exceptional sensitivity in the DC–MHz frequency range.

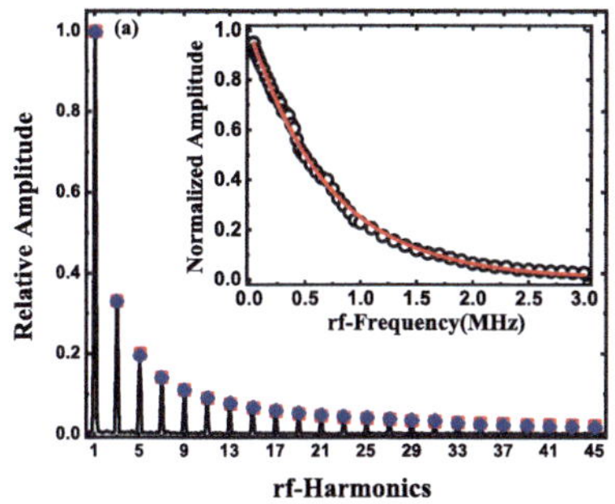

Fig. 6. Demonstration of rf magnetic-field spectrum analysis. The applied rf magnetic field is modulated by a square wave with a fundamental frequency of 10 kHz. The detected spectrum (solid black line) shows the fundamental and higher harmonics, normalized to the fundamental component. The red open squares represent the measured amplitudes of the harmonics, while the blue diamonds indicate the expected amplitudes for an ideal square wave. The harmonic amplitudes are further normalized by the sensor bandwidth, as shown in the inset. The open circles denote the experimental data, and the solid line represents an exponential fit.

Acknowledgments. Authors acknowledge Sushree S Sahoo, G. Rajalakshmi and Sujit Garain for their valuable contributions to the work. NM acknowledge financial support from i-Hub quantum technology foundation and SNS acknowledges the financial support from the DST-INSPIRE Faculty Fellowship grant No. DST/INSPIRE/04/2023/002297. The authors gratefully acknowledge the financial support from the Department of Atomic Energy, Government of India, under Project Identification No. RIN4001 (National Institute of Science Education and Research Bhubaneswar).

References

1. Ye, J., Zoller, P.: Essay: quantum sensing with atomic, molecular, and optical platforms for fundamental physics. Phys. Rev. Lett. **132**, 190001 (2024)
2. Fancher, C.T., Scherer, D.R., John, M.C.S., Marlow, B.L.S.: Rydberg atom electric field sensors for communications and sensing. IEEE Trans. Quantum Eng. **2**, 1–13 (2021), Art no. 3501313. https://doi.org/10.1109/TQE.2021.3065227
3. Mohapatra, A.K., Jackson, T.R., Adams, C.S.: Coherent optical detection of highly excited rydberg states using electromagnetically induced transparency. Phys. Rev. Lett. **98**, 113003 (2007)
4. Lin, Y.-Y., She, Z.-Y., Chen, Z.-W., Li, X.-Z., Zhang, C.-X., Liao, K.-Y., Zhang, X.-D., Chen, J.-H., Huang, W., Yan, H., Zhu, S.-L.: Terahertz receiver based on room-temperature Rydberg-atoms. Fundam. Res. **4**, 1154 (2023). https://doi.org/10.1016/j.fmre.2023.02.019
5. Legaie, R., Raithel, G., Anderson, D.: A millimeter-wave atomic receiver. AVS Quantum Sci. **6**, 024402 (2024). https://doi.org/10.1116/5.0173654
6. Cox, K.C., Meyer, D.H., Fatemi, F.K., Kunz, P.D.: Quantum-limited atomic receiver in the electrically small regime. Phys. Rev. Lett. **121**, 110502 (2018)

7. Zhao, B., Li, L., Zhang, Y., Tang, J., Liu, Y., Zhai, Y.: Optically pumped magnetometers recent advances and applications in biomagnetism: a review. IEEE Sens. J. **23**, 18949–18962 (2023)
8. Tierney, T.M., Holmes, N., Mellor, S., López, J.D., Roberts, G., Hill, R.M., Boto, E., Leggett, J., Shah, V., Brookes, M.J., Bowtell, R., Barnes, G.R.: Optically pumped magnetometers: from quantum origins to multi-channel magnetoencephalography. Neuroimage **199**, 598–608 (2019)
9. Romalis, M.V.: Optically pumped magnetometers for biomagnetic measurements. In: *Flexible High Performance Magnetic Field Sensors*, pp. 3–15. Springer, Cham (2022)
10. Kitching, J., Knappe, S., Shah, V., Schwindt, P., Griffith, G., Jimenez, R., Preusser, J., Liew, L.-A., Moreland, J.: Microfabricated atomic magnetometers and applications. In: 2008 IEEE International Frequency Control Symposium, pp. 789–794. IEEE (2008)
11. Aslam, N., Zhou, H., Urbach, E.K., Turner, M.J., Walsworth, R.L., Lukin, M.D., Park, H.: Quantum sensors for biomedical applications. Nat. Rev. Phys. **5**, 157–169 (2023)
12. Francis, T.J.V., Suna, R.R., Madhu, P.K., Viswanathan, N.K., Rajalakshmi, G.: Ultra-sensitive single-beam atom-optical magnetometer using weak measurement method. AIP Adv. **9**, 065113 (2019). https://doi.org/10.1063/1.5090581
13. Kurian, K.G., Sahoo, S.S., Madhu, P., et al.: Single-beam room-temperature atomic magnetometer with large bandwidth and dynamic range. Phys. Rev. Appl. **19**(5), 054040 (2023)
14. Budker, D., Jackson Kimball, D.F. (eds.): Optical Magnetometry. Cambridge University Press (2013)
15. Kominis, I.K., Kornack, T.W., Allred, J.C., Romalis, M.V.: A subfemtotesla multichannel atomic magnetometer. Nature **422**, 596–599 (2003)
16. Seltzer, S.J., Romalis, M.V.: Unshielded three-axis vector operation of a spin-exchange-relaxation-free atomic magnetometer. Appl. Phys. Lett. **85**, 4804–4806 (2004)
17. Keder, D.A., Prescott, D.W., Conovaloff, A.W., Sauer, K.L.: An unshielded radio-frequency atomic magnetometer with sub-femtotesla sensitivity. AIP Adv. **4**, 5449 (2014)
18. Rushton, L.M., Pyragius, T., Meraki, A., Elson, L., Jensen, K.: Unshielded portable optically pumped magnetometer for the remote detection of conductive objects using eddy current measurements. Rev. Sci. Instrum. **93**, 2402 (2022)
19. Sahoo, S.S., Mishra, S.R., Rajalakshmi, G., Mohapatra, A.K.: Nonlinear magnetoelectric effect in atomic vapor and its application to precision radio-frequency magnetometry. Phys. Rev. A **105**, 063509 (2022)
20. Romero-Abad, D.: The magnetic field in the axis of an elliptic loop current. Phys. Educ. **56**, 025014 (2021)

Quantum Communication Using Spatially Entangled Photons

Kiran Bajar, Vikas S. Bhat, and Sushil Mujumdar(✉)

Tata Institute of Fundamental Research, Mumbai, India
mujumdar@tifr.res.in|https://www.tifr.res.in/nomol

Abstract. In this presentation, we first lay down the basic essentials of Quantum Key Distribution (QKD), emphasising the source of unbreakable security embedded in the principles of quantum mechanics. We allude to the usage of entangled photons in QKD and elaborate on position-momentum (spatial) entanglement that we have achieved in our laboratory. Next, we describe our recent research on two topics, namely, the control of spatial entanglement in a photon beam, and the optimal segmentation of a two-dimensional detector to maximise the key rate in high-dimensional quantum key distribution. Finally, we speculate on what further QKD research can be undertaken collectively across various units of the Department of atomic energy (DAE).

Keywords: quantum key distribution · position-momentum entanglement · HDQKD

1 Introduction

Quantum communication is currently accepted as the most secure form of communication between a sender, typically known as Alice, and a receiver, usually called Bob in communication parlance. When Alice wants to send a message to Bob, she first encrypts the message, called plaintext, into unintelligible ciphertext. For that, she uses a key, which is essentially a random string of bits 0 and 1. In the quantum domain, these bits, now called qubits, are encoded using some physical attribute of quantum entities, such as for example, the polarization state of single photons. A vertically polarized photon can be labeled qubit 0, while a photon polarized at 45° to the vertical (henceforth called 45° photon, for brevity) can be qubit 1.[1] Thus, a string of photons with these polarization states will transfer the key from Alice to Bob. Once the key is successfully and confidentially transferred, the actual ciphertext can be relayed by any public channel. The security of the message is not compromised so long as the key remains uncracked. The security of the key lies in two principles of quantum mechanics, namely, the impossibility of duplication of quantum states, and the

[1] Orthogonally polarized photons are not usable in quantum communication to avoid deterministic measurement of both the qubits.

G. B. Mohanty and V. S. Prabhudesai (Eds.): DAEC 2026, SPPHY 368, pp. 49–58, 2026.
https://doi.org/10.1007/978-981-92-2948-2_7

probabilistic measurement of quantum superpositions. The no-cloning theorem [1] asserts that an arbitrary quantum state cannot be duplicated in a way where the original and the duplicate can co-exist, indeed, the original gets destroyed in the process of duplication. Next, the exact measurement of a superposition state cannot be made in a single measurement on a single entity. For instance, a 45° photon, that is an equal superpostion of 0° and 90°, when measured in a setup designed for 0° and 90°, is measured as either 0° or 90° with 0.5 probability. (Similarly, a 0° or a 90° photon in a setup made for 45° and 135° will have a similar consequence.) Since any hacking operation on a qubit involves its measurement and then its reproduction as suggested in the measurement, it is naturally subject to these measurement errors. Overall, these errors creep into the messaging protocol and form a red flag that gives away a hacker that is trying to eavesdrop into the communication.

An additional degree of security is imparted into the procedure when entangled photons are used. Photons that are entangled in polarization do not possess a defined polarization prior to measurement. Upon measurement of the polarization of one of an entangled pair, the polarization of the other is instantly fixed and shows up accordingly in a measurement. Communication protocols can be implemented on either a fiber-based platform or a free-space platform, or a combination of both. Both platforms have their pros and cons and the actual requirements of the situation suggests the right choice. For instance, a fiber depolarizes a photon over distances larger than about 150 km, and also, fiber infrastructure has to be laid down in advance, disallowing it from contingencies. It is also hard to set up on difficult terrain. On the other hand, free-space implementations are lossy, and have a very stringent pointing requirements. The most important experimental requirement is a reliable single photon source. Various sources are proposed and are regularly utilized. Commonly used on-demand sources involve quantum dots and color centers in various materials. A common example of the latter are nitrogen vacancy centers in diamond, which can absorb light and emit photons subsequently. On the other hand, a globally accepted bright source of single photons is a nonlinear process called spontaneous parametric downconversion (SPDC). This is a quantum optical process in which a single high-energy photon splits into two lower-energy photons when passing through a nonlinear optical crystal. It is one of the most common ways to generate entangled photon pairs for quantum optics and quantum information experiments. The splitting process happens probabilistically and very rarely (typically one pair per millions or billions of pump photons). This process is subject to energy conservation, and momentum must also be conserved through a condition called phase matching. Because crystals are birefringent, refractive indices depend on polarization direction. By carefully orienting the crystal, the pase matching condition can be satisfied. These conditions determine the emission angles, polarization, and wavelength distribution of outgoing photons. Often the photons emerge in cones around the pump beam. The photon pairs are created simultaneously with constraints such as single origin event, energy conservation, momentum conservation, and polarization correlations. however, since the crystal does not assign which photon gets which property, the pair is created in a quantum superposition, leading to entanglement in various properties such as polarization, position-momentum, time-energy, angle-orbital angular momentum etc. The correlation in position that occurs during the birth process (hence observed in the near-field), and the anti-correlation in momentum that manifests in the far-field is jointly called position-momentum entanglement or spatial entanglement.

2 Spatially Entangled Photons

2.1 Controlling the Degree of Spatial Entanglement

The field of two-photon interferometry was born out of some pioneering experiments done in the late 80s [2,3] and early 90s [4] when entangled photons generated out of SPDC were used. They resulted in a theoretical exposition by Horne et al. [5] giving a unified framework to describe these experiments. This further led to theoretical studies that examined the relationship between the one- and two-photon interference visibilities, through a beamsplitter, and through a double slit and established an inequality between one- and two-photon visibilities $V_m^2 + V_{12}^2 \leq 1$. Later, Saleh et al. expressed the equality case in terms of van Cittert–Zernike theorem and showed the analogy between partial coherence and partial entanglement, which they experimentally demonstrated for a double slit [6]. Their experiment records the interference fringes using an ICCD at various distances between the double slit and the detector. Since then, several experiments have probed various properties of the spatially entangled photons. It was used by Neves et al. to characterise the spatial entanglement present in the down-converted photons by measuring the Schmidt decomposition [7]. Soon, building on this work, Peeters et al. [8] showed a way to engineer the general 2-qubit state using a double slit by using imaging/Fourier optics after the double slit and by varying the slit separation. The multi-mode nature of the SPDC was exploited in a double slit experiment that violated the which-slit and interference inequality [9].

We explored the coherence present in the biphoton state and demonstrated this duality between the one- and two-photon visibilities by continuously tweaking the number of independent emitters that participate in generating the entangled photons. We did so by varying the size of the pump waist falling on our nonlinear crystal and quantifying the visibilities and their complementarity. We further report use cases for the various degrees of spatial entanglement that we generate and interpret our results using the theoretical treatment that we develop in terms of the concept of the birth zone number introduced by Schneeloch et al. [10]. We interpret the transfer of coherence from the individual photons to coherence between the entangled photons as a consequence of the changing birth zone number.

2.2 High-Dimensional Quantum Key Distribution: The Fryum Way

In today's information age, the exponential growth of global data transmission [11] has been coupled with an even faster growth in the security breaches of data

[11]. These data breaches are not just a concern for cyber-security experts but for the general public as well. With error-corrected quantum computing now being implemented on more and more qubits [12], our current encryption schemes are increasingly vulnerable to Shor's algorithm [13]. An inevitable alternative is information-theoretic secure encryption guaranteed by the laws of quantum mechanics, which have led to a few protocols. For instance, a quantum key distribution (QKD) protocol proposed by Brassard, Bennet and Mermin in 1992 called BBM92 [14] encodes a random binary key in the polarization of single photons. Once distributed, the two parties share a string of identical random bits that can be used as a key for secure communications. Although attractive for their unparalleled security, the actual implementation of these protocols using photon polarization is limited to a restrictive key generation rate of up to 0.5 bits/photon. This hinders the technology's widespread adaptability.

The widely accepted remedy towards this drawback is high-dimensional QKD (HDQKD) where the keys are encoded in higher dimensions. The benefit of implementing HDQKD is two-fold. First, the enhanced dimension size increases the bitrate and second, the inherent security and robustness to noise are enhanced [15,16]. For example, the generalization of BBM92 for d dimensions leads to a bitrate of $\log_2(d)/2$ per qudit and an error threshold for an optimal cloning attack of $(\sqrt{d}-1)/2\sqrt{d}$ [15]. These benefits have led to realizing various candidates for HDQKD using orbital angular momentum of entangled photons [17], time-bin qudits [18], transverse momentum [19] and polarization-momentum-time hybrid degrees of freedom [20]. These implementations, however, also come with certain demerits. The time-bin implementations suffer from high errors when scaled due to phase instability. The rest require specialized optical elements like a spatial-light modulator, a raster scanning pinhole, a digital micromirror device, a multicore fibre etc., that are either slow, lossy or hard to scale. Effectively, either the qudit rate or the dimension is constrained.

We maximized the key rate of a completely optically passive implementation of a position-momentum higher dimensional variant of the BBM92. The conjugate bases of position and momentum are easily realised using a single optical lens via imaging or Fourier transform. We used the underlying structure of the wavefunction of entangled photons generated by spontaneous parametric down-conversion (SPDC) to bifurcate the transverse beam's detection events into d independent and equiprobable segments. To form these segments, we grouped pixels on the detector into macropixels. We found the optimal segmentation by maximizing the total number of macropixels while simultaneously minimizing the error due to crosstalk between them. We showed that, counter-intuitively, discarding certain events leads to a gain in the secure key rate. We used this strategy to calculate the optimal segmentation, the fryum, for a set of wavefunction parameters that are available to us. Experimentally, we realise a secure key rate of 3.2 ± 0.1 bits per photon. Our method is only limited by the generation rate of entangled photons, the detection rate, and the dark counts, promising significantly higher bit rates (Fig. 1).

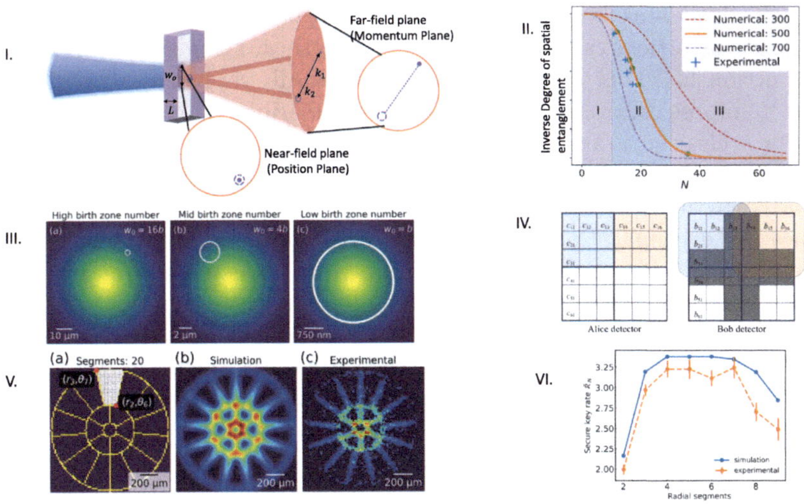

Fig. 1. I. Schematic of spontaneous parametric downconversion showing an incident high frequency pump beam downconverted to lower frequency two-photon beam with momentum correlations in far field and position correlations in near field. II. Inverse degree of spatial entanglement in a beam as a function of number of birth zones. III. Three representative images showing (a) beam with many birth zones (b) few birth zones, and (c) one birth zone. IV. Possible macropixelation on the Alice detector (left panel) for qudit-labelling, which results in grey-shaded ambiguous area on the Bob detector (right panel). V. A fryum-wheel segmentation (a) with 20 segments. (b) Simulation of spread of photon landing positions on Bob, when photons on Alice land on the segment boundaries (yellow lines in (a)). (c) Experimentally measured spread on Bob. VI. Secure key rate calculated from Shannon entropy as a function of number of radial segments in the fryum-wheel.

3 Wavefront Correction for Quantum Light

High-dimensional spatial entanglement of photon pairs offers substantial advantages over qubit-based platforms, including enhanced information capacity, improved noise resilience, and increased security in quantum communication and imaging [21,22]. However, these advantages are critically challenged when entangled photons propagate through complex or disordered media, which distort spatial correlations and degrade entanglement [23]. While the influence of disorder in the near field of a nonlinear crystal used for spontaneous parametric down-conversion (SPDC) was relatively well understood [24], the effect of far-field phase distortions on two-photon correlations remained less explored and posed a major obstacle for real-world implementations.

We first addressed this challenge at a fundamental level by systematically analyzing how arbitrary far-field phase distortions affected spatially entangled photon pairs. By decomposing random phase distortions into odd- and even-parity components, we theoretically demonstrated that two-photon correlations were exclusively sensitive to the even-parity part of the phase modulation, while odd-parity distortions canceled out due to the intrinsic symmetry of the biphoton wavefunction [25]. This prediction was experimentally verified using a deformable mirror to impose controlled phase aberrations described by Zernike polynomials, allowing a clear separation of odd and even parity contributions. The experimental observations showed excellent agreement with theory and were further supported by numerical simulations extending the results to stronger disorder regimes. A key implication of this finding was that the effective dimensionality of the disorder relevant to two-photon correlations was reduced by half, immediately suggesting a powerful simplification for adaptive correction strategies in quantum systems.

Building on this physical insight, we then translated the parity-based robustness of two-photon correlations into a practical wavefront correction strategy. We introduced a symmetrized genetic algorithm (sGA) that explicitly enforced even-parity constraints during optimization, ensuring that only physically relevant degrees of freedom were explored [26]. By reducing the optimization parameter space by a factor of two, sGA achieved faster convergence and higher enhancement within a finite number of generations compared to a conventional genetic algorithm (GA). Importantly, we experimentally and numerically established how the achievable enhancement depended on the signal-to-noise ratio of the feedback signal, which was governed by detector integration time in photon-counting experiments. This result highlighted a key trade-off between acquisition time and correction performance and demonstrated that parity-aware optimization was particularly advantageous in quantum-limited regimes. The sGA framework thus provided an efficient and scalable approach for restoring entanglement in quantum imaging, secure communication, and sensing through complex media.

Finally, recognizing that most quantum optical systems operated under severe photon constraints, we investigated the broader question of how optimization algorithms performed in the low-photon-counting regime. Through a systematic experimental and numerical comparison of two commonly used optimization algorithms—the genetic algorithm and the partitioning algorithm—across varying mean photon rates, we identified their complementary strengths and limitations under shot-noise-dominated conditions. Motivated by these insights, we proposed and implemented a novel augmented optimization algorithm that combined the robustness of one method with the rapid convergence of the other, making it particularly well suited for quantum applications [27]. This augmented approach demonstrated superior performance at low photon flux, where conventional algorithms often failed due to noisy feedback signals.

Together, these results established a unified framework for understanding and correcting entanglement degradation in complex environments. By connecting fundamental symmetry properties of biphoton wavefunctions with algorithmic design and photon-efficient optimization, this work provided both conceptual insight and practical tools for advancing high-dimensional quantum technologies under realistic experimental conditions.

Apart from these fundamental physics experiments described above, our lab is investing time and efforts in studying other exquisite states of quantum light, namely, 'phase-entangled states' and 'pure phase-entangled states'. Furthermore, experiments are also ongoing on a different domain of communication, namely, the communication between two quantum computers, something that is supposed to form the basis of the quantum internet. Here, we are studying the coupling of lanthanide ions to magnetic fields to pave the towards quantum transduction. On the technological front, we have initiated the development of realizing a free-space quantum link over a distance of about 1000 km, using a stratospheric balloon as a trusted node. A balloon floating at about 40 km altitude is expected to communicate in a line-of-sight manner with ground stations upto 500 km away. The corresponding pointing and stabilization requirements are currently being built.

4 Other Ongoing and Possible Activities in DAE

Finally, we discuss what we perceive as possible opportunities of DAE-wide collaborative efforts towards large-scale quantum communication initiatives. Apart from our research listed above, an experimental and theoretical effort is underway in IGCAR, Kalpakkam. The team is involved in QSDC, Quantm Secure Direct Communication [28], using polarization encoded single photons. The primary difference between QSDC and QKD is that the former implies sending the actual message and not a key. So there are no redundant requirements of two channels as in QKD, one for the message and the other of the key. In the protocol proposed and implemented by the team at IGCAR, the initiation is done by Bob by sending a photon with a polarization state written by Bob. This initialized photon reached Alice, who is supposed to let the photon through a quarter and a half wave plate aligned in a confidential configuration. This configuration essentially acts as a Unitary Transformation (UT) on the polarization state of the photon. This UT modifies the polarization state of the photon, which is then sent back to Bob. Bob, in turn, now sends this photon through his own UT, which again changes the state of polarization of the photon. Bob then analyzes this polarization through a detector setup. After analyzing two consecutive photons, Bob can unambiguously identify the confidential UT employed by Alice. The qubits are encoded in this UT and not in the polarization state of the photon. Since the photon is initialized by Bob himself, he can essentially keep track of every photon sent and received, and this ensures immunity to photon loss. Ultimately, the entire message is encoded in the series of UT employed by Alice and decoded by Bob. The protocol appears well-suited for fiber-based QSDC and ought to be field-tested over a local area network.

In a possible breakthrough experiment, the expertise of RRCAT in creating and maintaining a vacuum in long tubes can be exploited to build a high-dimensional QKD (HDQKD) experiment. In this experiment, the spatially-entangled photons discussed above are to be sent through a long evacuated tube through a carefully-designed series of lenses that ensures Fourier-transformed field distribution on a detection plane. This technique is called Vacuum Beam Guiding[29]. Depending on the dimensionality of the spatial entanglement, a lot denser information can be packed into one photon, thereby significantly increasing the qudit throughput. Toward this experiment, the expertise of the TIFR group in spatial entanglement and QKD can be combined with the expertise of the LIGO-India group in RRCAT in evacuated long tubes to implement this high-impact experiment, which promises to be the first-of-its kind in India, and depending on the distance covered, possibly in the world. The results can point to the possibility of a technological breakthrough in quantum key distribution efforts in the country and the world.

Finally, another relevant area of research for DAE could be that in quantum internet. Our group in TIFR is investing efforts in quantum transduction. Quantum transduction essentially involves writing the quantum state of a static qubit (superconducting Josephson junction, in our case) onto a photon, and back. This is aimed at networking quantum computers, which, by definition, work on superposition states. A variety of technologies have been proposed towards transduction, out of which our group is involved in rare-earth ions. These ions are sensitive to magnetic fields and the resulting Zeeman-splitting can admit microwave transitions that match the superconducting qubit frequencies. In that regard, the rare-earth elements researchers in DAE can contribute to this effort and strengthen the presence of quantum technologies in DAE.

Acknowledgments. We acknowledge the Department of Atomic Energy, Government of India, for funding for Project Identification No. RTI4002 under DAE OM No. 1303/1/2020/R&D-II/DAE/5567, Ministry of Science and Technology, India. We acknowledge helpful discussions with Dr Gururaj Kadiri, IGCAR, Kalpakkam, regarding their work on QSDC. We acknowledge helpful discussions with Dr Sendhil Raja, RRCAT, Indore, regarding their evacuated tube expertise towards Vacuum Beam Guiding.

Disclosure of Interests. The authors have no competing interests.

References

1. Wootters, W.K., Zurek, W.H.: A single quantum cannot be cloned. Nature **299**, 802–803 (1982)
2. Hong, C.K., Ou, Z.Y., Mandel, L.: Measurement of subpicosecond time intervals between two photons by interference. Phys. Rev. Lett. **59**, 2044–2046 (1987)
3. Ghosh, R., Mandel, L.: Observation of nonclassical effects in the interference of two photons. Phys. Rev. Lett. **59**, 1903–1905 (1987)
4. Zou, X.Y., Wang, L.J., Mandel, L.: Induced coherence and indistinguishability in optical interference. Phys. Rev. Lett. **67**, 318–321 (1991)
5. Horne, M.A., Shimony, A., Zeilinger, A.: Two-particle interferometry. Phys. Rev. Lett. **62**, 2209–2212 (1989)
6. Abouraddy, A.F., Nasr, M.B., Saleh, B.E.A., Sergienko, A.V., Teich, M.C.: Demonstration of the complementarity of one- and two-photon interference. Phys. Rev. A **63**, 063803 (2001)

7. Neves, L., Lima, G., Fonseca, E.J.S., Davidovich, L., Pádua, S.: Characterizing entanglement in qubits created with spatially correlated twin photons. Phys. Rev. A **76**, 032314 (2007)
8. Peeters, W.H., Renema, J.J., van Exter, M.P.: Engineering of two-photon spatial quantum correlations behind a double slit. Phys. Rev. A **79**, 043817 (2009)
9. Menzel, R., Heuer, A., Puhlmann, D., Dechoum, K., Hillery, M., Spähn, M., Schleich, W.: A two-photon double-slit experiment. J. Mod. Opt. **60**, 86–94 (2013)
10. Schneeloch, J., Howell, J.C.: Introduction to the transverse spatial correlations in spontaneous parametric down-conversion through the biphoton birth zone. J. Opt. **18**, 053501 (2016)
11. Cisco: Cisco Annual Internet Report—White Paper. https://www.cisco.com/c/en/us/solutions/collateral/executive-perspectives/annual-internet-report/white-paper-c11-741490.pdf (2020)
12. Radnaev, A.G., Chung, W.C., Cole, D.C., Mason, D., Ballance, T.G., Bedalov, M.J., Belknap, D.A., Berman, M.R., Blakely, M., Bloomfield, I.L., et al.: A Universal Neutral-Atom Quantum Computer with Individual Optical Addressing and Non-destructive Readout (2024)
13. Shor, P.W.: Polynomial-time algorithms for prime factorization and discrete logarithms on a quantum computer. SIAM J. Comput. **26**, 1484–1509 (1997)
14. Bennett, C.H., Brassard, G., Mermin, N.D.: Quantum cryptography without Bell's theorem. Phys. Rev. Lett. **68**, 557–559 (1992)
15. Cerf, N.J., Bourennane, M., Karlsson, A., Gisin, N.: Security of quantum key distribution using d-level systems. Phys. Rev. Lett. **88**, 127902 (2002)
16. Ecker, S., Bouchard, F., Bulla, L., Brandt, F., Kohout, O., Steinlechner, F., Fickler, R., Malik, M., Guryanova, Y., Ursin, R., Huber, M.: Overcoming noise in entanglement distribution. Phys. Rev. X **9**, 041042 (2019)
17. Mafu, M., Dudley, A., Goyal, S., Giovannini, D., McLaren, M., Padgett, M.J., Konrad, T., Petruccione, F., Lütkenhaus, N., Forbes, A.: Higher-dimensional orbital-angular-momentum-based quantum key distribution with mutually unbiased bases. Phys. Rev. A **88**, 032305 (2013)
18. Bechmann-Pasquinucci, H., Tittel, W.: Quantum cryptography using larger alphabets. Phys. Rev. A **61**, 062308 (2000)
19. Walborn, S.P., Lemelle, D.S., Almeida, M.P., Ribeiro, P.H.S.: Quantum key distribution with higher-order alphabets using spatially encoded qudits. Phys. Rev. Lett. **96**, 090501 (2006)
20. Ferro, L.F., Avelar, A.T., Cardoso, W.B., Gomes, R.M.: Single photon hybrid quantum key distribution. Phys. Scr. **99**, 025102 (2024)
21. Cerf, N.J., Bourennane, M., Karlsson, A., Gisin, N.: Security of quantum key distribution using d-level systems. Phys. Rev. Lett. **88** (2002)
22. Erhard, M., Krenn, M., Zeilinger, A.: Advances in high-dimensional quantum entanglement. Nat. Rev. Phys. **2**, 365–381 (2020)
23. Peeters, W.H., Moerman, J.J., Van Exter, M.P.: Observation of two-photon speckle patterns. Phys. Rev. Lett. **104**(17) (2010)
24. Lib, O., Hasson, G., Bromberg, Y.: Real-time shaping of entangled photons by classical control and feedback. Sci. Adv. 6(37) (2020)
25. Bajar, K., Shekel, R., Bhat, V.S., Chatterjee, R., Bromberg, Y., Mujumdar, S.: Rapid and Efficient Wavefront Correction for Spatially Entangled Photons Using Symmetrized Optimization (2025)
26. Bajar, K., Shekel, R., Bhat, V.S., Chatterjee, R., Bromberg, Y., Mujumdar, S.: Rapid and efficient wavefront correction for spatially entangled photons using symmetrized optimization. APL Photon. **10** (2025)

27. Bajar, K., Chatterjee, R., Bhat, V., Mujumdar, S.: Optimization of iterative wavefront shaping algorithms for quantum light. Opt. Lett. (2025)
28. Pan, D., Long, G.-L., Yin, L., Sheng, Y.-B., Ruan, D., Ng, S.X., Lu, J., Hanzo, L.: The evolution of quantum secure direct communication: on the road to the Qinternet. IEEE Commun. Surv. Tutor. **26**(3), 1898–1949 (2024)
29. Huang, Y., Salces-Carcoba, F., Adhikari, R.X., Safavi-Naeini, A.H., Jiang, L.: Vacuum beam guide for large scale quantum networks. Phys. Rev. Lett. **133**, 020801 (2024)

Quantum Technologies: a Short Note

Swati Choudhary[1,2], Sayan Mondal[1], Debarupa Saha[1], and Ujjwal Sen[1(✉)]

[1] Harish-Chandra Research Institute, A CI of Homi Bhabha National Institute, Chhatnag Road, Jhunsi, Prayagraj, India
ujjwal@hri.res.in

[2] Center for Quantum Science and Technology (CQST), Center for Computational Natural Sciences and Bioinformatics (CCNSB), International Institute of Information Technology Hyderabad, Prof. CR Rao Road, Gachibowli, Hyderabad, Telangana 500032, India

Abstract. This is a short review of the status of efforts to realize quantum devices, with the endeavors within India being focused—for the purpose of this note—on institutions within the Department of Atomic Energy of the Government of India.

1 Introduction

Rapid progress in quantum science and technology has led to the emergence of a diverse set of platforms that exploit uniquely quantum phenomena for practical tasks. In this short note, we present a concise overview of emerging quantum technologies, namely quantum communication networks [1–6], quantum computing [7–9], quantum sensing [10,11], and quantum simulation [12,13]. The development of these technologies can be traced back to the fundamental idea of simulating quantum systems using controllable quantum devices. This idea led to the formulation of quantum computers which utilize principles such as superposition and entanglement to perform certain computational tasks more efficiently than their classical counterparts. In parallel, the need for secure information transfer gave rise to quantum communication networks, where security is ensured through quantum features such as entanglement and nonorthogonality.

Quantum sensing has emerged as a powerful approach for achieving ultra-high precision measurements by exploiting quantum coherence and entanglement. In parallel, quantum simulators have been designed to model and investigate complex quantum systems that are otherwise intractable using classical methods.

We provide a brief discussion of these areas, along with indicators to the key institutes under the Department of Atomic Energy (DAE) of the Government of India that have contributed significantly to their development in recent years. The DAE forms an integral part of India's National Quantum Mission [14], a national initiative aimed at advancing research and technological capabilities in quantum science.

G. B. Mohanty and V. S. Prabhudesai (Eds.): DAEC 2026, SPPHY 368, pp. 59–74, 2026.
https://doi.org/10.1007/978-981-92-2948-2_8

DAE aims to advance quantum technologies across communication, sensing, and computation through a structured, long-term roadmap. In quantum communication, the goal is to realize secure long-distance networks using high-altitude platforms and satellite integration, ultimately enabling a national and global quantum internet. In parallel, DAE focuses on the development of quantum hardware, including photon sources, detectors, and large-area optical receivers, to support scalable quantum networks. In quantum sensing, the objective is to develop high-sensitivity devices such as superconducting quantum interference device (SQUID)-based magnetometers, terahertz (THz) sources and sensors, and high-resolution microscopy platforms. The roadmap emphasizes both fundamental studies, such as long-distance entanglement and quantum information processing under realistic conditions, and practical device development. A phased approach is envisioned, progressing from prototype development and experimental demonstrations to large-scale deployment and integration. In the long term, DAE aims at industrial-scale production and real-world applications in areas such as secure communication, healthcare, and astronomy. Overall, the vision is to establish indigenous capabilities and position India as a key contributor in next-generation quantum technologies.

The rest of the article is organized as follows:

Section 2 discusses quantum communication networks.

Section 3 presents a note on quantum computers.

Section 4 covers quantum sensors.

Section 5 addresses quantum simulators.

The note concludes with a brief summary in Sect. 6.

2 Quantum Communication Networks

A network is an assemblage of interconnected nodes that may be distributed over large geographical distances and are linked through transmission lines and communication channels to enable the exchange of information. The modern day internet is a quintessential example of such a network.

Conventional networks relies on the transmission of classical information encoded in the form of bits, which take values 0 or 1. These bits are processed, transmitted, and received across networks to enable various forms of communication and data exchange. Such bits are termed *classical* because they represent two distinct (distinguishable) and well-defined states of a system (for example, a ball can either be blue, denoted by 0, or red, denoted by 1).

In contrast, developments in quantum mechanics have revealed that particles such as electrons can exhibit behavior with no classical analog. In particular, quantum systems can exist in a superposition of distinct classical states. For instance, a quantum entity, e.g. a quantum bit ("qubit"), may be described by a state of the form $(|0\rangle + |1\rangle)/\sqrt{2}$, implying that it simultaneously possesses both "red" and "blue" characteristics.

Arguably, an even more remarkable feature of quantum systems is their ability to exhibit entanglement, wherein spatially separated particles exhibit correlations that have no classical counterpart. These intrinsically quantum traits have inspired the development of quantum networks; a large-scale interconnection of such networks is referred to as a quantum internet [1,5], aimed at harnessing quantum properties to achieve enhanced and qualitatively new information transmitting capabilities.

In essence, a quantum network enables the transmission of qubits across quantum channels [15], thereby facilitating the exchange of both classical (e.g., superdense coding [16]) and quantum information (e.g., quantum teleportation [17], which transfers quantum states using shared entanglement and classical communication). Quantum repeaters [18–20] serve as essential building blocks of such networks, aiding the transmission process by enabling the distribution of entanglement between distant nodes through entanglement swapping, entanglement purification, and the use of quantum memories.

Quantum networks offer several notable advantages over their classical counterparts. To begin with, one of the most crucial advantages of quantum networks lies in quantum key distribution (QKD) [21–23], which allows two distant parties to establish a shared secret key with security guaranteed by the laws of quantum mechanics. In contrast to classical cryptographic schemes, whose security relies on computational assumptions, quantum key distribution provides information-theoretic security. Any eavesdropping attempt inevitably disturbs the transmitted quantum states, essentially due to the no-cloning theorem [24,25], thereby revealing the presence of an adversary. Such advantages also extend to several other cryptographic primitives, including secret sharing [26–28], secure identification [29], and conference key agreement [30].

Such networks also open up new possibilities in distributed computation. By interconnecting smaller quantum processors, which are possibly entangled, one can potentially realize a modular architecture for quantum computing. Furthermore, such networks enable a user to delegate computations to a remote quantum processor while preserving the privacy of both the input and the computation [31].

In addition, quantum networks provide advantages in communication and information processing. The transmission of quantum states can reduce the amount of communication required for certain distributed tasks [32]. Moreover, quantum encoding and collective measurement strategies can enhance the rates of classical information transmission beyond what is achievable with purely classical resources [33–35].

Another benefit of quantum networks is that the distribution of entanglement across distant nodes enables significant improvements in quantum sensing [11, 36]. By correlating spatially separated systems, quantum networks can enhance precision in tasks such as time synchronization [37], surpassing classical limits.

Quantum correlations present in entangled states enable coordinated behavior between spatially separated nodes that cannot be reproduced using classical resources. Such coordination can be exploited in distributed tasks, including leader election [38] and achieving agreement among multiple parties. It can also provide advantages in cooperative scenarios, for instance by allowing spatially separated players in games such as bridge to coordinate their strategies more effectively [39].

Experimentally, significant progress has been made towards realizing quantum networks. Early large-scale implementations include the Defense Advanced Research Projects Agency (DARPA) quantum network in the United States, which pioneered multi-node QKD over metropolitan fiber links. In Europe, the Secure Communication based on Quantum Cryptography (SECOQC) network in Vienna and the SwissQuantum network in Geneva established secure communication using trusted-node architectures. Japan developed the Tokyo QKD network, which demonstrated secure real-time applications such as encrypted voice and video communication.

India has made significant progress in quantum communication. The Indian Space Research Organization (ISRO) has demonstrated free-space quantum communication over a distance of 300 m using indigenous technologies, including precise time-synchronization and optical alignment systems. Similar efforts are in progress in several other institutes and organizations including the Centre for Development of Telematics (C-DOT), Defense Research and Development Organization (DRDO), Indian Institute of Technology (IIT) Delhi, IIT Madras, Physical Research Laboratory (PRL), QNu Labs Pvt. Ltd., and Raman Research Institute (RRI). Such achievements represent important early steps toward satellite-based quantum communication, a key requirement for realizing long-distance and global-scale quantum networks. Other institutes involved in quantum communication experiments include Indian Institute of Science Education and Research (IISER) Mohali, IIT Kanpur, IIT Ropar, Indian Institute of Science (IISc), and Indian Institute of Space Science and Technology (IIST).

Several institutions under the Department of Atomic Energy are actively engaged in advancing quantum communication research. These include the Bhabha Atomic Research Centre (BARC), the Indira Gandhi Centre for Atomic Research (IGCAR), the Institute of Mathematical Sciences (IMSc), the Tata Institute of Fundamental Research (TIFR), and the Harish-Chandra Research Institute (HRI).

The DAE, in its “Amrit Kaal Vision Document” [40], proposes the development of long-distance quantum and free-space optical communication using indigenous platforms such as the TIFR Balloon Facility in Hyderabad, along with the design and realization of the required hardware. The plan includes establishing High-Altitude Platform (HAP) nodes for quantum communication over distances of 800–1000 km, with eventual integration with satellite-based systems. The initiative emphasizes secure communication through quantum technologies alongside high-speed optical communication, recognizing the overlap in their hardware and software requirements and the use of platforms such as satellites, drones, and aerostats. The implementation is planned in stages, beginning with the development of hardware, wide-area optical receivers, and HAP nodes, followed by demonstrations of networks of HAP nodes and large-area receivers. In 2025, researchers at TIFR reported a key milestone by demonstrating the preservation of entanglement between photon pairs over a distance of 100 m.

3 Quantum Computers

The concept of a quantum computer originated from the idea of simulating physical quantum systems that are intractable for conventional classical digital computers. This possibility was first proposed by Richard Feynman in 1980 [7], who suggested that such simulations could be efficiently performed using a quantum device, now known as a quantum simulator. It was then proposed that quantum devices called "quantum computers", which operate on qubits, and exploit superposition as a computational resource ("quantum parallelism"), can offer advantages in solving computational problems more efficiently than classical computers. This question was formulated by Deutsch [8], and was answered affirmatively in subsequent works [9,41,42], which demonstrated that quantum computers can achieve superpolynomial and, in some cases, exponential speedups over classical counterparts for certain problems. Shor [42] provided a polynomial-time algorithm for prime factorization and getting discrete logarithms on a quantum computer. Soon Grover's algorithm [43] was discovered, which enables searching an unstructured database of size N in $O(\sqrt{N})$ time, providing a quadratic speedup over the classical $O(N)$ requirement. These results demonstrated that quantum computers could solve certain problems more efficiently, thereby motivating the development of a complexity-theoretic framework to formalize such advantages. This led to the introduction of quantum complexity classes, such as BQP (Bounded-Error Quantum Polynomial Time), and to the study of their relationships with classical classes like P and NP. Consequently, quantum complexity theory [44] was established as a framework to characterize the capabilities and limitations of quantum computation.

Importantly, Shor's algorithm implies that the widely-used public-key cryptographic protocols are insecure if quantum computers become realizable, providing a fillip to the development of quantum cryptography [45–48] where security relies on the assumption that quantum mechanics is a correct description of natural events.

Notwithstanding the huge progress towards improved computational capabilities, one of the major hurdles in making this a reality is the inherent fragility of quantum behavior. Interactions with the environment induce decoherence in quantum systems, thereby hindering reliable computation [49]. This then led to the introduction of quantum error correcting codes [50–52] and fault tolerant methods [53,54]. Quantum error-correcting codes encode logical qubits into entangled states of multiple physical qubits to protect quantum information against noise and decoherence. They enable detection and correction of errors without directly measuring the encoded quantum state, thereby preserving coherence and enabling fault-tolerant computation. Motivated by these theoretical advances and growing curiosity for potentially more powerful computational machines, experimental efforts sought to bring these ideas into practice, leading to a range of proposals that demonstrated how platforms such as ion traps [55,56], cavity QED [57], and NMR (Nuclear Magnetic Resonance) techniques [58] can be harnessed to realize quantum logic operations.

Major organizations actively pursuing the development of fully functional quantum computers include Google, IBM, D-Wave, Quantinuum, IonQ, QuEra, and Rigetti. The Google team developed a programmable quantum processor, named Sycamore, based on superconducting circuit technology. This device consists of 53 operational qubits arranged in a two-dimensional lattice, that is able to implement entangling two-qubit gates between adjacent qubits. For more details, see [59]. In 2023, IBM demonstrated the real-time evolution of a two-dimensional transverse-field Ising model using its 127-qubit Eagle superconducting quantum processor [60].

In India, several organizations are advancing quantum computing technologies, including QNu Labs, Quanfluence, and C-DAC (Centre for Development of Advanced Computing).

Research groups in India working on ion-trap and photonic approaches to quantum computation include those at IISER Pune, IITs Delhi, Guwahati, Madras, and Roorkee, BARC, and RRI. Whereas superconducting qubit-based quantum computation is pursued at the IISc, TIFR (Mumbai, Hyderabad), and TCG-CREST, IISERs Mohali and Pune are actively involved in NMR-based quantum computing experiments. The aim of DAE [40] in this regard is to achieve an indigenous fault-tolerant quantum computer, focusing on superconducting qubits, ion-traps, and topological qubits. The aim is to develop around 10 logical qubits in near future, with increasing the capability to 100 logical qubits in 12–15 years. The long-term goal is to develop India into a major market and supplier of quantum computers.

4 Quantum Sensors

Quantum sensors are devices that utilize uniquely quantum features to estimate or detect physical parameters encoded in a system with high precision [11]. Quantum sensing typically involves preparing a system, referred to as the probe, in an initial quantum state, followed by encoding the parameter of interest, say θ, through a quantum evolution. This process imprints the parameter onto the quantum state. For a unitary evolution, the parameter is encoded as a relative phase, which can be then be extracted using interferometric protocols. The precision $\Delta\theta$ of such estimation schemes is fundamentally bounded by the Quantum Cramér-Rao bound, $\Delta\theta \geq 1/\sqrt{\nu F_Q}$, where ν is the number of repetitions and F_Q denotes the quantum Fisher information (QFI). For independent probes, the QFI scales linearly with the number of resources, $F_Q \sim N$, leading to the standard quantum limit $\Delta\theta \sim 1/\sqrt{N}$. In contrast, by utilizing quantum resources, it is possible to achieve $F_Q \sim N^2$, corresponding to the Heisenberg limit (HL) where $\Delta\theta \sim 1/N$ [10].

Beyond such approaches, an alternative is provided by critical quantum sensing [36, 61], where enhanced sensitivity arises from the susceptibility of many-body systems near a quantum phase transition. In this regime, small variations in a control parameter λ lead to large changes in the ground state, which are quantified by the QFI and related to the fidelity susceptibility χ_F as $F_Q \sim \chi_F$.

Near criticality, χ_F exhibits super-extensive scaling, often of the form $F_Q \sim N^\alpha$ with $\alpha > 1$, depending on the system. Consequently, critical systems can provide enhanced sensitivity.

Quantum sensors can be implemented in a wide variety of physical platforms, each offering distinct advantages. Atomic systems [62], including cold atoms and trapped ions, are widely used for high-precision measurements due to their excellent isolation and controllability. Solid-state systems, particularly defects in crystals such as nitrogen-vacancy centers [63, 64] in diamond, enable sensing at the nanoscale and under ambient conditions. Superconducting circuits provide strong coupling to microwave fields and are compatible with existing quantum computing architectures, while photonic and light-matter systems [65] leverage squeezed and entangled states of light for ultra-precise interferometry [66]. Nuclear magnetic resonance techniques also play an important role in sensitive detection of magnetic fields and have well-established applications in chemistry and medicine. More recently, hybrid systems combining different physical platforms have been explored to enhance sensitivity and enable new functionalities.

The applications of quantum sensing are broad and impactful. Atomic clocks [67, 68] represent one of the most successful examples, providing the most accurate time standards and forming the basis of global navigation systems. Atom-interferometry-based gravimeters and accelerometers are used in geophysical exploration, navigation, and tests of general relativity. At smaller scales, quantum thermometry [69] allows precise measurement of temperature in nanoscale systems. Quantum magnetometers [70, 71] can detect extremely weak magnetic fields, enabling various applications. Atomic magnetometers, for instance, use optically pumped atomic ensembles to achieve sensitivities comparable to or even exceeding superconducting devices. The SQUID [72, 73] is one of the most sensitive magnetometers available, relying on quantum interference of superconducting currents across Josephson junctions. SQUIDs can detect magnetic fields as small as femtotesla, making them essential in applications such as medical diagnostics (e.g., brain imaging), geophysical exploration, and fundamental physics experiments. Quantum-enhanced microscopy and imaging techniques leverage nonclassical states of light to surpass classical resolution and sensitivity limits. Quantum imaging methods improve sensitivity and allow imaging in low-light or noisy environments. These techniques are particularly promising for applications in biological imaging, materials science, and nanoscale characterization, where precision and minimal invasiveness are crucial.

Experiments in this direction are being carried out in several institutions in India including IIT Bombay (Qmet Tech Foundation), IISERs Bhopal and Kolkata, IISc, and RRI. Research on quantum clocks is being undertaken at the CSIR-National Physical Laboratory, DRDO, ISRO, IISER Pune, Inter-University Centre for Astronomy and Astrophysics (IUCAA), and RRI. The DAE in their Amrit Kaal Vision Document [40], has stated that it is of immense strategic and social importance that India develops its own optical clocks within the next decade and demonstrates operation at the level which is commensurate with global standards. Recently, researchers at the Raja Ramanna Centre for Advanced Technology (RRCAT) have developed a cold-atom based gravimeter [74]. Several other research groups across DAE institutions are engaged in a range of theoretical and experimental studies on quantum sensing and metrology, including investigations into quantum advantage and the development of realistic noise models. The long term goal is to have industrial scale production and deployment of such technologies.

5 Quantum Simulators

Simulating physical systems is a central task in science, as it enables both understanding of underlying mechanisms and prediction of future behavior. For a classical system, in the simplest case of a two-level system, only a single bit of information is required to specify its state. Consequently, a system composed of N such classical units can be described using N bits, implying that the required resources scale linearly with system size. In contrast, the state of an N-qubit quantum system is described by a (complex) vector in a 2^N-dimensional Hilbert space, requiring 2^N complex numbers (disregarding the normalization). Thus, the classical simulation of quantum systems generally demands resources that grow exponentially with N. This exponential overhead makes the simulation of large quantum systems intractable on classical devices. A promising approach to overcome this challenge is to employ well-controlled quantum systems to simulate other, less accessible quantum systems. Such devices are known as quantum simulators [12,13]. Quantum simulators hold promise for addressing a wide range of problems across various fields, including condensed-matter physics, high-energy physics, atomic physics, quantum chemistry, and cosmology. While quantum simulation can, in principle, be realized using universal quantum computers, it may also be implemented using simpler analog devices. Such analog quantum simulators typically require less precise control, making them comparatively easier to design and construct.

Quantum simulation involves evolving a system from an initial state $|\psi\rangle$ to a final state $|\psi'\rangle$ under a unitary evolution U. In a simulator, an initial state $|\phi\rangle$ is prepared and evolved using U_{sim} to a state $|\phi'\rangle$. If a suitable mapping between the system and the simulator exists, then the dynamics of the system can be reproduced using the simulator. Thus, the three main tasks in quantum simulation are state preparation, evolution, and measurement of the final state to extract the desired information. Traditionally, quantum simulators are classified as analog and digital. In digital quantum simulators (DQS), unitary operations are implemented using a set of universal quantum gates. These simulators are universal and highly programmable, with the potential for quantum error correction and precise control over dynamics. Their limitations include gate errors, noise accumulation, and, currently, limited qubit numbers. Analog quantum simulators (AQS), on the other hand, are physical systems engineered such that their Hamiltonian closely mimics the target Hamiltonian, allowing the system to naturally reproduce the desired quantum dynamics. Their main advantages are scalability to large system sizes and continuous-time evolution without the need for gate decomposition.

Any physical system that can be used as a quantum computer can also act as a platform for DQS [75], which includes superconducting qubit and trapped-ion platforms. There are many platforms for quantum simulation including atomic ensembles [76,77], NMR [78–80], superconducting circuits [81], and photonic systems [82]. There are various applications of quantum simulators in the areas of physics and chemistry. Quantum simulations tests various theoretical models that are not possible to be verified on classical simulators. Quantum simulators have been extensively used in solving quantum many-body problems. The Hubbard model, which is a model of interacting particles on a lattice has been simulated [83–85] on atomic and molecular systems, trapped in optical lattices, coupled cavities, ions, and superconducting circuits. Similarly, spin models have also been extensively studied using quantum simulators [86–88]. Quantum simulations of various novel quantum phenomena like superconductivity, quantum phase transitions, and topological orders have also been an area of active research. There are numerous problems in quantum chemistry that remain intractable for classical computation, like calculating reaction rates, modeling catalysis, electron transport, and bond energies and bulk properties in moderately-sized molecules. Quantum simulators are increasingly being applied in the area of high-energy physics and cosmology as well. Quantum simulations of systems relevant to high-energy physics [89,90] like gauge theories, and nucleons have been proposed in ion-trap and neutral atom setups, as well as in DQS. Quantum simulations are also used to study analog gravity and cosmological models [91,92] in trapped-ions and ultra-cold atom simulators.

AQS has emerged as a particularly successful approach over the past decade. Notably, large-scale quantum spin models have been experimentally realized on programmable platforms with up to 256 qubits [93], demonstrating the scalability and versatility of analog devices. In contrast, digital quantum simulation (DQS) is currently limited by noise and decoherence in available hardware. Present-day Noisy Intermediate-Scale Quantum (NISQ) devices nevertheless offer a viable platform for quantum simulation tasks [75]. In this regime, variational quantum algorithms (hybrid classical-quantum methods) are especially promising. These algorithms rely on relatively shallow quantum circuits, with the computationally intensive optimization performed on classical processors, making them well-suited for near-term devices. In quantum chemistry [94], Variational Quantum Eigensolvers (VQEs) have been used to simulate small molecular systems [95,96]. Industrial efforts in quantum simulation have expanded rapidly, encompassing neutral-atom platforms (QuEra Computing, Pasqal), trapped-ion systems (IonQ, Quantinuum), superconducting qubits (IBM, D-Wave Systems, Google Quantum AI), and photonic architectures (Xanadu).

In India, several research groups are actively engaged in both theoretical and experimental studies of quantum simulation. Experimental efforts on trapped-ion platforms are being pursued at institutions such as IISER Pune and IIT Kanpur. In parallel, neutral-atom-based experiments are being carried out at

IITs Guwahati and Roorkee, RRI, IISER Pune, IISc, and NISER, among others. The DAE units working on quantum simulation include IMSc, Institute of Physics (IOP), NISER, Saha Institute of Nuclear Physics (SINP), TIFR, and HRI.

6 Conclusion

In this short note, we have discussed the current landscape of quantum technologies, encompassing quantum communication, computation, sensing, and simulation, with particular emphasis on developments within the DAE institutes. These domains, while distinct in their objectives, are deeply interconnected through the common framework of quantum mechanics and collectively contribute to the broader vision of next-generation technologies. Quantum communication promises qualitatively better classical and quantum information transfer which can be fundamentally secure, quantum computation offers computational advantages in classical problems including for classically intractable problems, quantum sensing enables ultra-precise measurements beyond classical limits, and quantum simulation provides a powerful approach to understanding complex quantum systems that are otherwise inaccessible.

In quantum communication, extending distances and developing efficient quantum repeaters are crucial. Quantum computing requires advances in error correction, fault tolerance, and hardware scalability. Quantum sensing must address robustness and real-world deployability, while quantum simulators need improved control, verification techniques, and integration with computational frameworks.

Looking ahead, progress in quantum technologies will likely rely on hybrid and interdisciplinary approaches, combining platforms such as photonic, atomic, and solid-state systems to enhance performance. Advances in error mitigation, scalable architectures, and entanglement distribution will be key to realizing practical quantum computing and large-scale quantum networks. At the same time, developments in many-body physics and variational methods are expected to further strengthen quantum sensing and simulation. The DAE aims to become a major supplier and market for various quantum technologies. To achieve this end, it is crucial to forge academic-industry partnership in the research and development of quantum technologies.

In conclusion, quantum technologies are transitioning from foundational research to early-stage applications. While significant challenges remain, continued theoretical insights, experimental breakthroughs, engineering leaps, and coordinated efforts are expected to transform these technologies into practical tools with far-reaching scientific and societal impact.

Acknowledgments. We acknowledge discussions with members of the HRI Quantum Technologies Group, in particular, with Sayan Choudhury, Abhishodh Prakash, Debraj Rakshit, and Aditi Sen De. We also acknowledge useful exchanges with participants at the 2nd DAE Conclave (DAEC2026) at TIFR Mumbai.

References

1. Wehner, S., Elkouss, D., Hanson, R.: Quantum internet: a vision for the road ahead. Science **362**, eaam9288 (2018). https://www.science.org/doi/10.1126/science.aam9288
2. Singh, A., Dev, K., Siljak, H., Joshi, H.D., Magarini, M.: Quantum internet—applications, functionalities, enabling technologies, challenges, and research directions. IEEE Commun. Surv. Tutor. **23**, 2218 (2021). https://ieeexplore.ieee.org/abstract/document/9528843
3. Zhang, P., Chen, N., Shen, S., Yu, S., Wu, S., Kumar, N.: Future quantum communications and networking: a review and vision. IEEE Wirel. Commun. **31**, 141 (2022). https://ieeexplore.ieee.org/abstract/document/9928082
4. Wei, S.-H., Jing, B., Zhang, X.-Y., Liao, J.-Y., Yuan, C.-Z., Fan, B.-Y., Lyu, C., Zhou, D.-L., Wang, Y., Deng, G.-W., et al.: Towards real-world quantum networks: a review. Laser Photonics Rev. **16**, 2100219 (2022). https://doi.org/10.1002/lpor.202100219
5. Azuma, K., Economou, S.E., Elkouss, D., Hilaire, P., Jiang, L., Lo, H.-K., Tzitrin, I.: Quantum repeaters: from quantum networks to the quantum internet. Rev. Mod. Phys. **95**, 045006 (2023). https://link.aps.org/doi/10.1103/RevModPhys.95.045006
6. Caleffi, M., Amoretti, M., Ferrari, D., Illiano, J., Manzalini, A., Cacciapuoti, A.S.: Distributed quantum computing: a survey. Comput. Netw. **254**, 110672 (2024). https://www.sciencedirect.com/science/article/pii/S1389128624005048
7. Feynman, R.P.: Simulating physics with computers. Int. J. Theor. Phys. **21**, 467–488 (1982). https://doi.org/10.1007/BF02650179
8. Deutsch, D.: Quantum theory, the church-turing principle and the universal quantum computer. Proc. Roy. Soc. Lond. A **400**, 97–117 (1985). https://doi.org/10.1098/rspa.1985.0070
9. Simon, D.R.: On the power of quantum computation. In: Proceedings 35th Annual Symposium on Foundations of Computer Science, pp. 116–123. (1994). https://doi.org/10.1109/SFCS.1994.365701
10. Giovannetti, V., Lloyd, S., Maccone, L.: Quantum-enhanced measurements: beating the standard quantum limit. Science **306**, 1330–1336 (2004). https://www.science.org/doi/abs/10.1126/science.1104149
11. Degen, C.L., Reinhard, F., Cappellaro, P.: Quantum sensing. Rev. Mod. Phys. **89**, 035002 (2017). https://doi.org/10.1103/RevModPhys.89.035002
12. Georgescu, I.M., Ashhab, S., Nori, F.: Quantum simulation. Rev. Mod. Phys. **86**, 153–185 (2014). https://doi.org/10.1103/revmodphys.86.153
13. Altman, E., Brown, K.R., Carleo, G., et al.: Quantum simulators: architectures and opportunities. PRX Quantum **2**, 017003 (2021). https://doi.org/10.1103/PRXQuantum.2.017003
14. Department of Science and Technology, Government of India: National Quantum Mission (NQM). https://dst.gov.in/national-quantum-mission-nqm
15. Van Meter, R.: Quantum Networking. Wiley (2014)
16. Bennett, C.H., Wiesner, S.J.: Communication via one- and two-particle operators on Einstein-Podolsky-Rosen states. Phys. Rev. Lett. **69**, 2881 (1992). https://link.aps.org/doi/10.1103/PhysRevLett.69.2881
17. Bennett, C.H., Brassard, G., Crépeau, C., Jozsa, R., Peres, A., Wootters, W.K.: Teleporting an unknown quantum state via dual classical and einstein-podolsky-rosen channels. Phys. Rev. Lett. **70**, 1895 (1993). https://link.aps.org/doi/10.1103/PhysRevLett.70.1895

18. Dür, W., Briegel, H.-J., Cirac, J.I., Zoller, P.: Quantum repeaters based on entanglement purification. Phys. Rev. A **59**, 169 (1999). https://doi.org/10.1103/PhysRevA.59.169
19. Sangouard, N., Simon, C., de Riedmatten, H., Gisin, N.: Quantum repeaters based on atomic ensembles and linear optics. Rev. Mod. Phys. **83**, 33 (2011). https://link.aps.org/doi/10.1103/RevModPhys.83.33
20. Munro, W.J., Azuma, K., Tamaki, K., Nemoto, K.: Inside quantum repeaters. IEEE J. Sel. Top. Quantum Electron. **21**, 78 (2015). https://ieeexplore.ieee.org/abstract/document/7010905
21. Bennett, C.H., Brassard, G.: Quantum cryptography: public key distribution and coin tossing. Theor. Comput. Sci. **560**, 7 (2014). https://www.sciencedirect.com/science/article/pii/S0304397514004241
22. Ekert, A.K.: Quantum cryptography based on Bell's theorem. Phys. Rev. Lett. **67**, 661 (1991). https://link.aps.org/doi/10.1103/PhysRevLett.67.661
23. Bennett, C.H.: Quantum cryptography using any two nonorthogonal states. Phys. Rev. Lett. **68**, 3121–3124 (1992). https://link.aps.org/doi/10.1103/PhysRevLett.68.3121
24. Wootters, W.K., Zurek, W.H.: A single quantum cannot be cloned. Nature **299**, 802 (1982). https://www.nature.com/articles/299802a0
25. Dieks, D.: Communication by EPR devices. Phys. Lett. A **92**, 271 (1982). https://doi.org/10.1016/0375-9601(82)90084-6
26. Żukowski, M., Zeilinger, A., Horne, M.A., Weinfurter, H.: Quest for GHZ states. Acta Phys. Pol., A **93**, 187 (1998)
27. Cleve, R., Gottesman, D., Lo, H.-K.: How to share a quantum secret. Phys. Rev. Lett. **83**, 648 (1999). https://link.aps.org/doi/10.1103/PhysRevLett.83.648
28. Hillery, M., Bužek, V., Berthiaume, A.: Quantum secret sharing. Phys. Rev. A **59**, 1829 (1999). https://doi.org/10.1103/PhysRevA.59.1829
29. Damgård, I.B., Fehr, S., Salvail, L., Schaffner, C.: Secure identification and qkd in the bounded-quantum-storage model. In: Annual International Cryptology Conference, p. 342. Springer (2007). https://doi.org/10.1016/j.tcs.2014.09.014
30. Murta, G., Grasselli, F., Kampermann, H., Bruß, D.: Quantum conference key agreement: a review. Adv. Quantum Technol. **3**, 2000025 (2020). https://advanced.onlinelibrary.wiley.com/doi/full/10.1002/qute.202000025
31. Broadbent, A., Fitzsimons, J., Kashefi, E.: Universal blind quantum computation. In: 2009 50th Annual IEEE Symposium on Foundations of Computer Science, p. 517. IEEE (2009)
32. Brassard, G.: Quantum communication complexity. Found. Phys. **33**, 1593 (2003). https://link.springer.com/article/10.1023/A:1026009100467
33. Sasaki, M., Kato, K., Izutsu, M., Hirota, O.: Quantum channels showing superadditivity in classical capacity. Phys. Rev. A **58**, 146 (1998). https://doi.org/10.1103/PhysRevA.58.146
34. Hastings, M.B.: Superadditivity of communication capacity using entangled inputs. Nat. Phys. **5**, 255 (2009). https://www.nature.com/articles/nphys1224
35. Guha, S.: Structured optical receivers to attain superadditive capacity and the holevo limit. Phys. Rev. Lett. **106**, 240502 (2011). https://link.aps.org/doi/10.1103/PhysRevLett.106.240502
36. Agarwal, K.D., Mondal, S., Sahoo, A., Rakshit, D., Sen, A., Sen, U.: Quantum sensing with ultracold simulators in lattice and ensemble systems: a review. Int. J. Mod. Phys. C. 2543006 (2025). https://www.worldscientific.com/doi/full/10.1142/S0129183125430065

37. Komar, P., Kessler, E.M., Bishof, M., Jiang, L., Sørensen, A.S., Ye, J., Lukin, M.D.: A quantum network of clocks. Nat. Phys. **10**, 582 (2014). https://www.nature.com/articles/nphys3000
38. Tani, S., Kobayashi, H., Matsumoto, K.: Exact quantum algorithms for the leader election problem. ACM Trans. Comput. Theory **4**, 1 (2012). https://dl.acm.org/doi/10.1145/2141938.2141939
39. Muhammad, S., Tavakoli, A., Kurant, M., Pawłowski, Ma., Żukowski, M., Bourennane, M.: Quantum bidding in bridge. Phys. Rev. X **4**, 021047 (2014). https://link.aps.org/doi/10.1103/PhysRevX.4.021047
40. Department of Atomic Energy, India: Amrit Kaal Vision Documents (2025). https://digilib.dae.gov.in/books/10736
41. Deutsch, D., Jozsa, R.: Rapid solution of problems by quantum computation. Proc. Roy. Soc. Lond. Ser. A: Math. Phys. Sci. **439**, 553–558 (1992). https://doi.org/10.1098/rspa.1992.0167
42. Shor, P.W.: Algorithms for quantum computation: discrete logarithms and factoring. In: Proceedings 35th Annual Symposium on Foundations of Computer Science, pp. 124–134. (1994). https://doi.org/10.1109/SFCS.1994.365700
43. Grover, L.K.: A Fast Quantum Mechanical Algorithm for Database Search. arXiv:quant-ph/9605043 [quant-ph]. https://arxiv.org/abs/quant-ph/9605043 (1996)
44. Bernstein, E., Vazirani, U.: Quantum complexity theory. SIAM J. Comput. **26**, 1411–1473 (1997). https://doi.org/10.1137/S0097539796300921
45. Gisin, N., Ribordy, G., Tittel, W., Zbinden, H.: Quantum cryptography. Rev. Mod. Phys. **74**, 145–195 (2002). https://doi.org/10.1103/RevModPhys.74.145
46. Scarani, V., Bechmann-Pasquinucci, H., Cerf, N.J., Dušek, M., Lütkenhaus, N., Peev, M.: The security of practical quantum key distribution. Rev. Mod. Phys. **81**, 1301–1350 (2009). https://doi.org/10.1103/RevModPhys.81.1301
47. Pirandola, S., Andersen, U.L.: Advances in quantum cryptography. Adv. Opt. Photon. **12**, 1012 (2020). https://doi.org/10.1364/aop.361502
48. Xu, F., Ma, X., Zhang, Q., Lo, H.-K., Pan, J.-W.: Secure quantum key distribution with realistic devices. Rev. Mod. Phys. **92**, 025002 (2020). https://doi.org/10.1103/RevModPhys.92.025002
49. Haroche, S., Raimond, J.-M.: Quantum computing: dream or nightmare? Phys. Today **49**, 51–52 (1996). https://doi.org/10.1063/1.881512
50. Peter, W.S.: Scheme for reducing decoherence in quantum computer memory. Phys. Rev. A **52**, R2493–R2496 (1995). https://doi.org/10.1103/PhysRevA.52.R2493
51. Steane, A.M.: Error correcting codes in quantum theory. Phys. Rev. Lett. **77**, 793–797 (1996). https://doi.org/10.1103/PhysRevLett.77.793
52. Kitaev, A.Y.: Quantum computations: algorithms and error correction. Russ. Math. Surv. **52**, 1191–1249 (1997). https://doi.org/10.1070/RM1997v052n06ABEH002155
53. Shor, P.W.: Fault-tolerant quantum computation. In: Proceedings of 37th Conference on Foundations of Computer Science, pp. 56–65 (1996). https://api.semanticscholar.org/CorpusID:7508572
54. Preskill, J.: Fault-Tolerant Quantum Computation. arXiv:quant-ph/9712048 [quant-ph] (1997). https://arxiv.org/abs/quant-ph/9712048
55. Cirac, J.I., Zoller, P.: Quantum computations with cold trapped ions. Phys. Rev. Lett. **74**, 4091–4094 (1995). https://doi.org/10.1103/PhysRevLett.74.4091
56. Monroe, C., Meekhof, D.M., King, B.E., Itano, W.M., Wineland, D.J.: Demonstration of a fundamental quantum logic gate. Phys. Rev. Lett. **75**, 4714–4717 (1995). https://doi.org/10.1103/PhysRevLett.75.4714

57. van Enk, S.J., Cirac, J.I., Zoller, P.: Purifying two-bit quantum gates and joint measurements in cavity qed. Phys. Rev. Lett. **79**, 5178–5181 (1997). https://doi.org/10.1103/PhysRevLett.79.5178
58. Vandersypen, L.M.K., Chuang, I.L.: NMR techniques for quantum control and computation. Rev. Mod. Phys. **76**, 1037–1069 (2005). https://doi.org/10.1103/RevModPhys.76.1037
59. Arute, F., Arya, K., Babbush, R., et al.: Quantum supremacy using a programmable superconducting processor. Nature **574**, 505–510 (2019). https://doi.org/10.1038/s41586-019-1666-5
60. Kim, Y., Eddins, A., Anand, S., et al.: Evidence for the utility of quantum computing before fault tolerance. Nature **618**, 500–505 (2023). https://doi.org/10.1038/s41586-023-06096-3
61. Montenegro, V., Mukhopadhyay, C., Yousefjani, R., Sarkar, S., Mishra, U., Paris, M.G.A., Bayat, A.: Review: quantum metrology and sensing with many-body systems. Phys. Rep. **1134**, 1–62 (2025). https://doi.org/10.1016/j.physrep.2025.05.005
62. Pezzè, L., Smerzi, A., Oberthaler, M.K., Schmied, R., Treutlein, P.: Quantum metrology with nonclassical states of atomic ensembles. Rev. Mod. Phys. **90**, 035005 (2018). https://doi.org/10.1103/RevModPhys.90.035005
63. Barry, J.F., Schloss, J.M., Bauch, E., Turner, M.J., Hart, C.A., Pham, L.M., Walsworth, R.L.: Sensitivity optimization for nv-diamond magnetometry. Rev. Mod. Phys. **92**, 015004 (2020). https://doi.org/10.1103/RevModPhys.92.015004
64. Du, J., Shi, F., Kong, X., Jelezko, F., Wrachtrup, J.: Single-molecule scale magnetic resonance spectroscopy using quantum diamond sensors. Rev. Mod. Phys. **96**, 025001 (2024). https://doi.org/10.1103/RevModPhys.96.025001
65. Forn-Díaz, P., Lamata, L., Rico, E., Kono, J., Solano, E.: Ultrastrong coupling regimes of light-matter interaction. Rev. Mod. Phys. **91**, 025005 (2019). https://doi.org/10.1103/RevModPhys.91.025005
66. Cronin, A.D., Schmiedmayer, J., Pritchard, D.E.: Optics and interferometry with atoms and molecules. Rev. Mod. Phys. **81**, 1051–1129 (2009). https://doi.org/10.1103/RevModPhys.81.1051
67. Martin, M.J., Bishof, M., Swallows, M.D., Zhang, X., Benko, C., von Stecher, J., Gorshkov, A.V., Rey, A.M., Ye, J.: A quantum many-body spin system in an optical lattice clock. Science **341**, 632–636 (2013). https://doi.org/10.1126/science.1236929
68. André, A., Sørensen, A.S., Lukin, M.D.: Stability of atomic clocks based on entangled atoms. Phys. Rev. Lett. **92**, 230801 (2004). https://doi.org/10.1103/PhysRevLett.92.230801
69. Mehboudi, M., Sanpera, A., Correa, L.A.: Thermometry in the quantum regime: recent theoretical progress. J. Phys. A: Math. Theor. **52**, 303001 (2019). https://doi.org/10.1088/1751-8121/ab2828
70. Vengalattore, M., Higbie, J.M., Leslie, S.R., Guzman, J., Sadler, L.E., Stamper-Kurn, D.M.: High-resolution magnetometry with a Spinor Bose-Einstein condensate. Phys. Rev. Lett. **98**, 200801 (2007). https://doi.org/10.1103/PhysRevLett.98.200801
71. Greenberg, Y.S.: Application of superconducting quantum interference devices to nuclear magnetic resonance. Rev. Mod. Phys. **70**, 175–222 (1998). https://doi.org/10.1103/RevModPhys.70.175
72. Kirtley, J.R., Ketchen, M.B., Stawiasz, K.G., Sun, J.Z., Gallagher, W.J., Blanton, S.H., Wind, S.J.: High-resolution scanning squid microscope. Appl. Phys. Lett. **66**, 1138–1140 (1995). https://doi.org/10.1063/1.113838

73. Bending, S.J.: Local magnetic probes of superconductors. Adv. Phys. **48**, 449 (1999). https://doi.org/10.1080/000187399243437
74. Bhardwaj, K., Singh, S., Ram, S.P., Jain, B., Kumar, V., Pathak, A., Tiwari, S., Tiwari, V.B., Mishra, S.R.: Measurement of gravitational acceleration in a single laser operated atomic fountain. arXiv preprint arXiv:2405.04223 (2024)
75. Fauseweh, B.: Quantum many-body simulations on digital quantum computers: state-of-the-art and future challenges. Nat. Commun. **15**, 2123 (2024). https://doi.org/10.1038/s41467-024-46402-9
76. Bloch, I., Dalibard, J., Nascimbène, S.: Quantum simulations with ultracold quantum gases. Nat. Phys. **8**, 267 (2012). https://doi.org/10.1038/nphys2259
77. Blatt, R., Roos, C.F.: Quantum simulations with trapped ions. Nat. Phys. **8**, 277–284 (2012). https://doi.org/10.1038/nphys2252
78. Peng, X., Du, J., Suter, D.: Quantum phase transition of ground-state entanglement in a Heisenberg spin chain simulated in an NMR quantum computer. Phys. Rev. A **71**, 012307 (2005). https://doi.org/10.1103/PhysRevA.71.012307
79. Peng, X.-H., Suter, D.: Spin qubits for quantum simulations. Front. Phys. China **5**, 1–25 (2009). https://doi.org/10.1007/s11467-009-0067-x
80. Peng, X., Zhang, J., Du, J., Suter, D.: Quantum simulation of a system with competing two- and three-body interactions. Phys. Rev. Lett. **103**, 140501 (2009). https://doi.org/10.1103/PhysRevLett.103.140501
81. Houck, A.H., Türeci, H.E., Koch, J.: On-chip quantum simulation with superconducting circuits. Nat. Phys. **8**, 292 (2012). https://doi.org/10.1038/nphys2251
82. Aspuru-Guzik, A., Walther, P.: Photonic quantum simulators. Nat. Phys. **8**, 285 (2012). https://doi.org/10.1038/nphys2253
83. Jaksch, D., Zoller, P.: The cold atom hubbard toolbox. Ann. Phys. **315**, 52–79 (2005). https://doi.org/10.1016/j.aop.2004.09.010. Special Issue
84. Greiner, M., Mandel, O., Esslinger, T., Hänsch, T.W., Bloch, I.: Quantum phase transition from a superfluid to a mott insulator in a gas of ultracold atoms. Nature **415**, 39 (2002). https://doi.org/10.1038/415039a
85. Deng, X.-L., Porras, D., Cirac, J.I.: Quantum phases of interacting phonons in ion traps. Phys. Rev. A **77**, 033403 (2008). https://doi.org/10.1103/PhysRevA.77.033403
86. Deng, X.-L., Porras, D., Cirac, J.I.: Effective spin quantum phases in systems of trapped ions. Phys. Rev. A **72**, 063407 (2005). https://doi.org/10.1103/PhysRevA.72.063407
87. Friedenauer, A., Schmitz, H., Glueckert, J.T., Porras, D., Schaetz, T.: Simulating a quantum magnet with trapped ions. Nat. Phys. **4**, 757 (2008). https://doi.org/10.1038/nphys1032
88. Edwards, E.E., Korenblit, S., Kim, K., Islam, R., Chang, M.-S., Freericks, J.K., Lin, G.-D., Duan, L.-M., Monroe, C.: Quantum simulation and phase diagram of the transverse-field Ising model with three atomic spins. Phys. Rev. B **82**, 060412 (2010). https://doi.org/10.1103/PhysRevB.82.060412
89. Martinez, E.A., Muschik, C.A., Schindler, P., Nigg, D., Erhard, A., Heyl, M., Hauke, P., Dalmonte, M., Monz, T., Zoller, P., Blatt, R.: Real-time dynamics of lattice gauge theories with a few-qubit quantum computer. Nature **534**, 516 (2016). https://doi.org/10.1038/nature18318
90. Bauer, C.W., Davoudi, Z., Baha Balantekin, A., Bhattacharya, T.: Quantum simulation for high-energy physics. PRX Quantum 4, (2023). https://doi.org/10.1103/PRXQuantum.4.027001

91. Fischer, U.R., Schützhold, R.: Quantum simulation of cosmic inflation in two-component Bose-Einstein condensates. Phys. Rev. A **70**, 063615 (2004). https://doi.org/10.1103/PhysRevA.70.063615
92. Menicucci, N.C., Olson, S.J., Milburn, G.J.: Simulating quantum effects of cosmological expansion using a static ion trap. New J. Phys. **12**, 095019 (2010). https://doi.org/10.1088/1367-2630/12/9/095019
93. Ebadi, S., Wang, T.T., Levine, H., et al.: Quantum phases of matter on a 256-atom programmable quantum simulator. Nature **595**, 227 (2021). https://doi.org/10.1038/s41586-021-03582-4
94. Cao, Y., Romero, J., Olson, J.P., et al.: Quantum chemistry in the age of quantum computing. Chem. Rev. **119**, 10856 (2019). https://doi.org/10.1021/acs.chemrev.8b00803
95. Kandala, A., Mezzacapo, A., Temme, K., Takita, M., Brink, M., Chow, J.M., Gambetta, J.M.: Hardware-efficient variational quantum eigensolver for small molecules and quantum magnets. Nature **549**, 242 (2017). https://doi.org/10.1038/nature23879
96. Arute, F., Arya, K., Babbush, R., Bacon, D., et al.: Hartree-fock on a superconducting qubit quantum computer. Science **369**, 1084 (2020). https://doi.org/10.1126/science.abb9811

Next-generation Materials

BARC Prospective in Changing Role of Materials for Advanced Nuclear Reactors in India

Raghvendra Tewari(✉)

Materials Group, Bhabha Atomic Research Centre, Mumbai 400085, India
raghvendra.tewari@gmail.com

Abstract. Challenges for materials scientists working in the field of nuclear energy is to continuously develop materials with superior properties as the nuclear industry demands newer materials that perform better than the current materials. The need for materials to function in a new environment is unprecedented as BARC is focussing on the development of Gen IV reactors. In this paper a few of the research topics, like rare earth and critical materials, materials for Gen IV reactors and materials for society, which are currently being pursued in Materials Group BARC, are presented.

Keywords: Rare Earth elements · Critical materials · Gen IV reactors · SiC · Nuclear grade graphite

1 Introduction

Since the inception of BARC, Materials Group (MG), together with other groups in BARC, has played a key role in developing a range of materials that Department of Atomic Energy (DAE) needs for its nuclear program as well as for other uses. As a consequence of this, activities of MG are spanning across entire material cycle. For example, Mineral Processing Division (MinD) along with AMD, UCIL and IREL works in identifying and qualifying minerals of interest of DAE. Detailed optimized flow sheets for economically extracting materials are being developed by MinD, Materials processing Division (MPD) and Powder Metallurgy Division (PMD). In addition, components of reactors are being developed, qualified and their constant performance evaluation being done by Materials Science (MSD) and Mechanical Metallurgy Divisions (MMD). UED and PMD also run small plants to produce specific materials which are constantly required by DAE. New compositions for hosting radioactive waste are being developed by GAMD and MSD to fix nuclear waste.

In initial days, the main work of MG was to ensure supply of uranium from indigenous sources. For this purpose, in 1955, a processing plant for extracting uranium from Monazite was established in BARC [1]. Crude uranium tetrafluoride taken out as byproduct was the raw material for the production of uranium needed for research reactors. Operational experiences of this plant led to the development of solvent extraction equipment,

G. B. Mohanty and V. S. Prabhudesai (Eds.): DAEC 2026, SPPHY 368, pp. 77–81, 2026.
https://doi.org/10.1007/978-981-92-2948-2_9

Magnesio-thermic reduction, ventilation systems of radioactive handling units, waste recycling, industrial scale chemical and instrumental analytical techniques.

Early experiences gained in this plant came very handy in the separation of chemically nearly identical rare-earth (RE) elements. Solvent extraction process, which was developed in the early stages of Uranium extraction, was extended in separation of oxides of RE elements. Tight control over chemical composition led to the separation of induvial oxides. These technologies were later provided to IREL for commercial exploitation of these oxides.

2 Rare Earth and Critical Materials

In the regard to rare earth elements, it is worth mentioning here that the term *rare-earth* is a mis-norm, as these elements are much more abundantly present in comparison to commonly known elements like silver. In fact, India ranks 5th in terms of reserves of RE based minerals with total deposit of ~ 69 lakh tons of RE oxides [2]. However, most of the RE deposits in India contain lesser important light RE elements like, La, Ce, Pr, Nd, (~94%) and technologically important heavy RE elements are sparsely present. Therefore, to process Indian deposits special care should be taken, which has been successfully achieved in BARC.

In addition to the separation of individual RE oxides, considerable efforts have been put on converting oxides into metallic forms. For this purpose, three techniques, namely; reduction-diffusion process, modified AMES process and molten salt electrolysis process have been developed. Using reduction diffusion process, a plant making Sm-Co based magnets has already been set up. For Nd metal production, modified AMES process has been employed and technology has been transferred to successfully produce Nd metal at industrial scale. Beside this, other possible sources of RE have been explored and extracting RE elements from Coal fly ash -for the first time in the world- has been initiated with Neyveli Lignite India Limited at Cuddalore, Tamil Nādu. Similarly, for the first time in India, a pilot plant is being set up for the extraction of Nd element from a source other than monazite.

The Secondary sources for other critical metals like Ni, Co, W, Mo, B, which do not have any primary sources in the country and are needed for strategic and non-strategic purposes, have been identified. Specific extraction flow sheet for each induvial element has been developed. For example, scrap of maraging steels produced in MIDHANI have been used to produce Ni, Co and Mo and with the help of Heavy water board, a demonstration plant has been successfully set up. Scraps of tool steel or WO_3 produced as by product has been used to produce W.

3 Materials for Gen IV Reactors

As BARC is moving towards development of Gen IV reactors, MG geared itself for the new challenges. The proposed Molten Salt Reactors use fuel in liquid state which is a composite fluoride salt at elevated temperatures of 500–650 °C [3]. As a result, structural material for such reactor has to perform under aggressive corrosive environment of fluorine ions, radiation doses and should have sufficient strength at elevated temperature to resist creep and other high temperature deformation mechanisms. From the view point of alloy design, it is a very challenging task as the alloy must concurrently satisfy two paradoxical conditions. While corrosion demands purest form of material to avoid formation of any galvanic cell, demand of high temperature strength, on the other hand, requires multi-component alloy system. Therefore, after series of experiments and detailed literature review, an optimized composition was determined. Experimental trails of the samples have shown that the alloy met all the required mechanical properties and met all qualifying criteria. Based on this qualification, full scale alloy development of 3-ton material was taken up and entire flowsheet of the forming and shaping in form of components for an experimental loop simulating reactor condition has been successfully achieved.

In Gen IV reactor, graphite is an essential material but for this purpose nuclear grade graphite is needed. The typical parameters of nuclear graphite should include isotropy better than 1.1, density exceeding 1.8 g/cc, and a B-equivalent less than 2 ppm. Meeting all the three requirements for a typical graphite material obtained from any natural resources are difficult to attain. Therefore, in place of modifying the conventional graphite, a completely new material called meso-carbon microbeads have been taken as a starting material and completely new processing parameters were developed to produce nuclear grade graphite. The process has been scaled up to the commercial scale and slabs of 0.2 m × 0.2 m × 0.4 mm are being successfully produced.

SiC is another material which is required for Gen IV reactors. For Gas cooled reactors (GCR) it is a candidate material for fuel tube applications. As the proposed coolant for GCR is CO_2 gas, the structural materials, apart from high temperature strength, should be compatible with high temperature CO_2 gas which becomes corrosive at elevated temperatures. In fact, CO_2 undergoes decomposition process at the surface of fuel tubes depositing C on the surface of fuel tubes. This gradually increases carbon content of fuel tube material. For SiC based material two parallel approaches have been adopted. In first approach, particulate composite of SiC based materials and in second approach SiC fibre based SiC composite material are being developed. In the later approach, as India does not have any facility for make SiC fibre, technology for making high quality SiC fibres were generated first and subsequently SiC fibre and SiC matrix-based composites are being prepared at Lab scale. The development work on the composite material of particulate SiC has achieved relatively better progress as nearly 97% of density has been successfully achieved. Structural stability of SiC under irradiation was examined by exposing samples to I^{7+} ions of 00 meV energy to a fluence of 7×10^{13} ions/cm^2 and no change in the structure was noticed. Joining of SiC components through a glass ceramic joint has been successfully achieved.

Reactor pressure vessel (RPV) is a heart of any PWRs. With demand of smaller and compact reactors, like small modular reactors (SMR), need of better materials is inevitable. In MG, a new class of bainitic steel has been developed. While conventional bainitic steels derive their strengths from the carbide phases which are unstable under irradiation environment, new generation steel avoids carbide phases and derives its strength from fine grain size to ~ 100 nm. Such steel has shown a strength at least 3 time of the conventional steel, while maintaining other properties similar to that of conventional bainitic steels.

4 Materials for Society

MG has developed several materials which find applications in the medical field. Recently developed Yttria loaded glass microsphere (Bhabhasphere) for Hepatocellular Carcinoma radiotherapy. These microspheres are 10–15 times cheaper than imported materials and have shown encouraging results when implemented in patients [4]. Recently Ho doped microspheres have been developed which are sensitive to magnetic field as well as give soft γ-rays which have propensity for external examination in comparison to Y-based materials where after operation in-vivo examination is necessary.

Carbon nanotubes have extremely high surface area, very responsive to change in surface condition and high selectivity to organic compounds [5]. These properties have been utilized for early-stage diagnostic purposes for cancers in different internal organs. For this, a patient needs to blow in a tube and all volatile compounds evolving from various cancer cells are collected over a template made out of carbon nanotube and change of voltage response is measured through systematic variation and coding. One such non-invasive device can find out cancer in any internal organ of human body. A prototype device has been successfully developed and few more models are being prepared for generating necessary data.

5 Epilogue

The role of advanced materials is rapidly evolving from enabling performance improvements to driving transformative innovations across industries. These materials are active contributors to sustainability, energy storage, and intelligent systems. As science converges with advanced manufacturing, advanced materials will increasingly shape the future of nuclear technology, economy, and society, positioning themselves as the backbone of next-generation solutions.

Acknowledgments. This work is collective efforts of various scientists working in MG. I am particularly thankful to all young colleagues of the group.

References

1. Sahu, M.L., Satpati, S.K., Chowdhury, S.: Uranium Metal Production Facility in BARC, Trombay, Mumbai. In: Proceedings of the Theme Meeting on the Journey of BARC Safety Council for Strengthening Safety Culture in BARC Facilities: 2000–2017, pp. 79–83 (2017)
2. Balaram, V.: Rare earth element deposits: sources, and exploration strategies. J. Geol. Soc. India **98**(9), 1210–1216 (2022)
3. Serp, J., Allibert, M., Beneš, O., Delpech, S., Feynberg, O., Getta, V., Heuer, D., Holcomb, D., Ignatiev, V., Kloosterman, J., Luzzi, L., Merle-Lucotte, E., Uhlíř, J., Yoshioka, R., Zhimin, D.: The molten salt reactor (MSR) in generation IV: overview and perspectives. Prog. Nucl. Energy **77**, 308–319 (2014)
4. Jha, A.K., Puranik, A.D., Gala, K.B., Nair, K.V., Rajeswari, A., Lohar, S.P., Shah, A., Mathai, A.M., Tiwary, B.K., Shettye, N.S., Mathur, A.: First-in-India experience with 90Y-BhabhaSphere: an indigenous 90Y-TheraSphere biosimilar for transarterial radioembolization of hepatic malignancies. Br. J. Radiol. **98**(1176), 2135–2143 (2025)

5. Ji, S., Liu, C., Zhang, B., Yang, F., Xu, J., Long, J., Jin, C., Fu, D., Ni, Q., Yu, X.: Carbon nanotubes in cancer diagnosis and therapy. Biochimica et Biophysica Acta (BBA)-Rev. Cancer **1806**(1), 29–35 (2010)

Quantum Materials for Spintronics: A Pathway to Energy-Efficient Computer Memory Technology

Subhankar Bedanta(✉) and Abhisek Mishra

Laboratory for Nanomagnetism and Magnetic Materials, School of Physical Sciences, National Institute of Science Education and Research (NISER), An OCC of Homi Bhabha National Institute (HBNI), Jatni 752050, India
sbedanta@niser.ac.in

Abstract. The global transition towards sustainable energy sources has significantly addressed challenges in energy generation. However, major inefficiencies persist at the level of energy utilization, particularly in digital technologies where waste heat from computation, data storage, and communication increasingly offsets the benefits of clean electricity. In this context, spintronics has emerged as a promising route toward energy efficient electronics by exploiting the electron's spin degree of freedom, in addition to its charge, to reduce Joule heating and overcome scaling limitations inherent to conventional charge based systems. Recent advances in quantum materials have further accelerated the development of spintronics by enabling efficient generation, manipulation, and detection of spin currents. These quantum materials provide new mechanisms for spin transport and magnetization control that are fundamentally distinct from traditional ferromagnetic metal systems. We discuss the evolution of spintronics from the Nobel Prize winning discovery of giant magnetoresistance, which revolutionized magnetic data storage, to modern spin orbit torque driven switching enabled by quantum materials. We discuss the historical progression of memory technologies, from early magnetic disk storage in the 1950s to contemporary spin orbit torque based magnetic random access memory, highlighting key advantages such as non-volatility, high endurance, low standby power, and potential as a universal memory. Beyond memory applications, quantum material-based spintronics enables a broad range of functionalities, including spin-based logic, neuromorphic computing, spin-torque oscillators, quantum spin qubits, and flexible spintronic devices. By integrating spin orbit phenomena with emerging quantum materials, spintronics offers a compelling pathway toward energy efficient computer memory technologies and the development of beyond CMOS computing paradigms, paving the way for next-generation sustainable information processing systems.

1 Introduction

The global energy landscape is undergoing a profound transformation as societies increasingly turn toward sustainable and low-carbon energy sources such

G. B. Mohanty and V. S. Prabhudesai (Eds.): DAEC 2026, SPPHY 368, pp. 82–98, 2026.
https://doi.org/10.1007/978-981-92-2948-2_10

as solar energy, hydropower, nuclear energy and wind energy to meet growing demands [1–3]. Hydropower, with its long established infrastructure, continues to provide reliable large-scale electricity generation with minimal direct emissions, while solar and wind energy have rapidly expanded over the past decade, driven by technological advances, declining costs, and strong policy support across the globe. Nuclear energy, owing to its exceptionally high energy density and stable capability, remains a crucial component of the energy for many countries seeking long-term energy security with low greenhouse gas emissions [1]. Together, these energy sources are reshaping how electricity is produced and distributed worldwide. However, the transition towards cleaner energy generation addresses only one side of the sustainability challenge. The other, equally critical, aspect lies in how efficiently this energy is utilized once it is generated. A substantial portion of the electricity produced, irrespective of its source, is ultimately lost as waste heat in electronic devices, computing hardware, communication networks, and data centers [4]. As modern society becomes increasingly digital, with explosive growth in artificial intelligence, high-performance computing, and cloud services, the energy consumption associated with information processing and data storage is rising at an unprecedented rate [5,6]. The challenge, therefore, is not merely to generate cleaner electricity but to minimize the energy required to perform computation, store information, and transmit data. Addressing this challenge requires a drastic shift from incremental engineering optimizations towards new physical principles that enable low-energy operation at the device level. Consequently, research efforts are increasingly focused on alternative information carriers and mechanisms that can reduce Joule heating and overcome the scaling limitations of conventional charge-based electronics. It is within this broader context of global energy efficiency, rather than energy generation alone, that spintronics has emerged as a compelling and timely approach for next-generation information technologies. Within this context, spintronics has emerged as a promising avenue towards energy-efficient electronics by exploiting the spin degree of freedom of electrons in addition to their charge [7–9]. The technological relevance of spintronics was firmly established with the discovery of Giant Magnetoresistance (GMR), which revolutionized magnetic data storage and earned Peter Grünberg and Albert Fert the Nobel Prize in Physics in 2007 [10]. GMR-based read heads enabled dramatic increases in storage density in hard disk drives, transforming computers from specialized laboratory instruments into compact, affordable, and widespread consumer usages. This breakthrough highlighted the power of magnetic thin films and positioned spin-dependent transport as a cornerstone of modern information technology.

Early data storage technologies were dominated by the Hard Disk Drive (HDD), where information is stored magnetically and written using external magnetic fields. While HDDs provided large storage capacity at low cost, they suffered from mechanical latency, high power consumption, and limited speed. To overcome these limitations, solid-state memories such as Solid-State Drive (SSD) emerged, based on NAND Flash Memory. SSDs significantly improved speed, durability, and energy efficiency by eliminating moving parts. However,

SSDs introduced new challenges, including limited write endurance, higher cost per bit, and scaling limitations due to charge leakage and reliability issues at smaller dimensions. As a result, most household computers continue to rely primarily on HDDs. In present day HDDs, the writing process relies on magnetic fields, which are energetically costly and become increasingly inefficient as device dimensions are scaled down. This imbalance, where reading data is energy-efficient but writing data requires large power due to magnetic field generation, has become a major obstacle in developing low-power and scalable memory technologies. Spintronics provides a promising solution to this challenge. Mechanisms such as spin-transfer torque (STT) and spin–orbit torque (SOT) allow magnetization switching to be achieved through spin currents instead of relying on external magnetic fields [11,12]. In particular, SOT-based devices are highly attractive because they decouple the write and read operations, that is, spin currents generated via the Rashba Edelstein (REE) and spin Hall Effect (SHE) are used for writing information, while magnetoresistive effects such as GMR or tunneling magnetoresistance (TMR) are employed for readout. This separation not only reduces power consumption but also improves switching speed, endurance, and device reliability.

Recent advances in materials science have extensively explored conventional high spin–orbit coupling (SOC) materials, such as platinum (Pt), for SOT-based devices. These heavy metals have been widely employed for generating spin currents via the spin Hall effect owing to their strong SOC [13]. However, their practical implementation faces several limitations, including high cost, limited material availability, and challenges in large-scale industrial integration, which restrict their economic viability for widespread memory technologies. In this context, emerging quantum materials including topological insulators, transition metal dichalcogenides, antiferromagnets, altermagnets, complex oxides etc have further advanced the field by enabling highly efficient spin orbit torque generation, improved spin transport, and reduced switching currents. These materials exhibit novel physical mechanisms such as spin momentum locking, large spin Hall angles, and unconventional magnetic ordering, making them highly promising candidates for next generation energy efficient spintronic memory devices. These advances have further contributed to the development of a new technology named, magnetic random access memory (MRAM), a non-volatile memory technology that combines high speed, excellent endurance, and low energy operation. In comparison to dynamic random access memory (DRAM), MRAM offers non-volatility, virtually unlimited write cycles, and significantly lower standby power, while continuing to improve in density and scalability. Although challenges related to cost and large-scale integration remain, MRAM is widely regarded as a strong contender for next-generation memory and even universal memory applications. Motivated by these considerations, this article provides a comprehensive overview of the evolution of spintronics from its foundational discoveries to contemporary device implementations, focusing on spin orbit driven mechanisms, magnetic memory technologies such as MRAM, and the opportunities offered by quantum materials for overcoming current energy and scaling bottlenecks in information technology.

2 Evolution of Memory Devices: From Magnetic Disks to Spintronic Memories

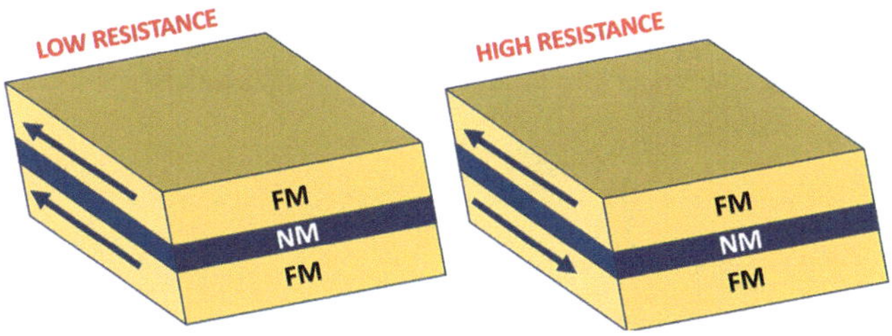

Fig. 1. Giant magnetoresistance in FM/NM/FM trilayer structure: schematic illustration of a ferromagnet/non-magnet/ferromagnet (FM/NM/FM) multilayer showing resistance dependence on magnetization alignment. A parallel alignment of the magnetizations in the two ferromagnetic layers (left) reduces spin-dependent scattering, leading to a lower resistance state. When the magnetizations are antiparallel (right), spin-dependent scattering increases, leading to high resistance.

The history of computing is inseparable from the evolution of memory technologies. While advances in processing power often dominate discussions of technological progress, it is the development of reliable, fast, and scalable memory that has ultimately enabled modern computing systems. Over the past decades, memory devices have evolved through multiple paradigms, that is, mechanical, magnetic, electronic, and now spintronic each driven by the same fundamental demands, which are, higher speed, greater density, lower energy consumption, and improved reliability [14]. The first major milestone in commercial data storage was the hard drive in IBM 305 Random Access Method of Accounting and Control (RAMAC), introduced in 1956 [15]. It was the world's first commercial computer to employ a hard disk drive for random-access storage. The RAMAC system used fifty 24-inch magnetic disks, rotating at high speed, to store approximately 5 MB of data an extraordinary achievement at that time. The RAMAC disk drive comprised 50 magnetically coated metal platters with a total storage capacity of approximately five million characters.

Despite its groundbreaking nature, the RAMAC was massive, expensive, and energy-intensive. Its storage density was extremely low by modern standards, and mechanical motion dominated read–write operations. Nevertheless, it established magnetic storage as a practical and scalable solution, setting the stage for decades of innovation in hard disk technology. Over the following decades, hard disk drives (HDDs) underwent continuous miniaturization and performance

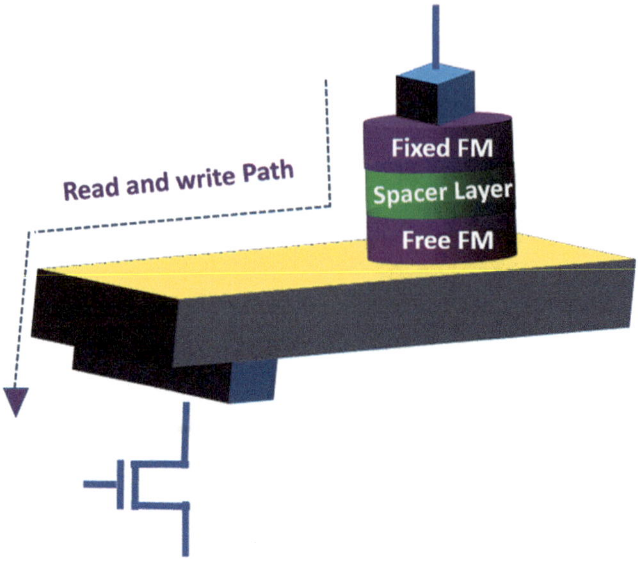

Fig. 2. Schematic illustration of an STT device consisting of a fixed ferromagnetic layer, non-magnetic spacer layer, and free ferromagnetic layer. A spin-polarized current flowing through the stack exerts spin-transfer torque on the free layer, enabling magnetization switching.

improvements. By late 2025 and early 2026, the maximum storage capacity available in a single HDD reached 36 TB with Seagate's Exos M (ST36000NM003K), marking the company's largest hard drive to date [16]. This release follows the 32 TB model introduced in late 2024, which itself succeeded the earlier 30 TB drive launched roughly a year prior, reflecting the continued progression in HDD storage density. It is worth noting that conventional HDDs retained inherent limitations. Writing data required magnetic fields generated by current-carrying coils, a process that consumed significant energy and scaled poorly with shrinking bit size. Mechanical motion limited access speed and introduced latency, reliability concerns, and power overheads. As computing workloads shifted toward high-speed, data-intensive applications, these limitations became increasingly pronounced. To overcome the limitations associated with mechanical motion and magnetic-field-based writing in HDDs, Solid-State Drives (SSD) emerged as a promising alternative. Unlike HDDs, SSDs rely on NAND flash memory to store information electronically, eliminating moving parts and significantly improving access speed, reliability, and energy efficiency. This solid-state architecture enables faster data retrieval, reduced latency, and enhanced durability, making SSDs particularly attractive for modern computing environments that demand high performance.

Early commercial solid-state drives (SSDs) emerged in the 1970s for applications requiring high speed and robust operation [17]. These early systems utilized volatile RAM chips, supported by batteries and hard disk drives to preserve data during power loss. In 1978, StorageTek (Louisville, Colorado) introduced the STC 4305 enterprise SSD, which offered a storage capacity of 45 MB at a cost of approximately $400,000 [17]. By the end of 2025, the largest available SSD reached a storage capacity of approximately 245.88 TB, developed by KIOXIA [18] . However, despite these advantages, SSDs also face limitations such as finite write endurance, charge leakage at smaller dimensions, and relatively higher cost per bit compared to HDDs. These challenges, along with the growing demand for high-speed and energy-efficient memory technologies, have motivated the exploration of alternative approaches beyond conventional charge-based storage.

While HDDs and SSDs are primarily designed for long term data storage, modern computing systems also require high speed memory for real time data processing and execution. Volatile semiconductor memories, including Static Random Access Memory (SRAM) and Dynamic Random Access Memory (DRAM), fulfill this requirement and constitute the upper levels of the memory hierarchy. SRAM, typically used in processor caches, stores information using bistable flip-flop circuits made of transistors and offers extremely fast access speeds and low latency but suffers from low density and high cost, restricting its use to small capacity applications. Unlike SRAM, DRAM stores data as charge in a capacitor, enabling higher density and reduced cost, thereby making it ideal for main system memory. However, periodic refresh operations are necessary to preserve the stored information, resulting in increased power consumption. Unlike HDDs and SSDs, both SRAM and DRAM are volatile and lose stored information when power is removed. These limitations highlight the fundamental trade-off between speed, power consumption, and data retention in conventional memory technologies, motivating the search for alternative solutions that combine high speed, non-volatility, and energy efficiency. These limitations of conventional memory technologies have motivated the exploration of alternative approaches that combine the speed of volatile memories with the non-volatility of storage devices. Spintronics has emerged as a promising solution in this direction by utilizing the electron's spin degree of freedom to enable fast, energy-efficient, and non-volatile memory technologies. Spintronics, or spin electronics, is a field of electronics that exploits not only the charge of electrons but also their spin, a fundamental quantum property associated with a magnetic moment.

In conventional electronics, information is processed using the flow of electrical charge, whereas spintronics exploits the electron's spin state (spin-up or spin-down) to store and manipulate information, enabling faster, energy-efficient, and non-volatile devices. A major breakthrough in spintronics was the discovery of Giant Magnetoresistance (GMR), which enabled highly sensitive magnetic read heads and significantly increased hard disk storage density. GMR occurs in multilayer structures of alternating ferromagnetic and nonmagnetic layers, where electrical resistance depends on the relative alignment of magnetizations as shown in Fig. 1. The resistance is high for antiparallel alignment and low for parallel alignment due to spin-dependent electron scattering [10]. This large resistance change under small magnetic fields makes GMR highly effective for magnetic readout applications.

MRAM represents a fundamental shift in memory technology by combining magnetic storage principles with semiconductor-style access speeds [10]. In MRAM, information is stored in the magnetic orientation of nanoscale ferromagnets, typically within magnetic tunnel junctions (MTJs). Data retention does not rely on stored charge, making MRAM inherently non-volatile. The key advantages of MRAM stem from this physical mechanism. MRAM offers non-volatility, allowing data to be retained without power, significantly reducing standby energy consumption. It also provides fast read and write speeds comparable to SRAM, low latency, and extremely high endurance, with very high write cycles. These attributes position MRAM as a strong candidate for replacing or complementing existing memory hierarchies. In this context, spintronic memories based on spin-transfer torque (STT) and spin–orbit torque (SOT) have emerged as promising candidates [19]. Unlike charge-based memories, STT-based magnetic random access memory (STT-MRAM) utilizes spin-polarized currents to switch magnetization states, enabling non-volatile operation, high endurance, and fast switching speeds (Fig. 2). In STT mechanism, the magnetization of a ferromagnetic material is switched using a spin-polarized current rather than an external magnetic field.

When an electric current passes through a magnetic layer, the electron spins become polarized along the magnetization direction of that layer. As this spin-polarized current enters another magnetic layer with a different magnetization orientation, the transfer of spin angular momentum exerts a torque on the local magnetic moments. If the applied current is sufficiently large, this torque can switch the magnetization direction between parallel and antiparallel states, enabling information writing in magnetic memory devices.Several companies, including Samsung, Everspin, Avalanche Technologies, IBM, Spin Transfer Technologies, and Crocus Technology, are actively developing STT-MRAM chips. Everspin announced the shipment of 256 Mb STT-MRAM samples to customers in April 2016 [65]. These devices demonstrated interface speeds comparable to DRAM, supporting DDR3 and DDR4 interfaces. In March 2025, Everspin introduced new automotive-grade STT-MRAM devices, EM064LX HR and EM128LX HR, with capacities up to approximately 128 Mb [20]. These devices operate over a wide temperature range from -40 to $+125\,^{\circ}$C, comply with the AEC-Q100 automotive standard, and are designed for applications in automotive, aerospace, defense, and industrial systems. However, STT devices typically require relatively high current densities ($\sim 10^6$ to 10^7 A/cm^2) and involve current flow directly through the tunnel barrier, which can impact device reliability and scalability.

To address these challenges, spin–orbit torque (SOT)-based memory has gained increasing attention, where spin currents generated in adjacent materials through spin–orbit coupling are used to switch magnetization efficiently [14,19]. By separating the read and write paths, SOT-based devices offer improved endurance, faster switching, and reduced power consumption, making them highly attractive for next-generation energy-efficient memory technologies. SOT is a mechanism for switching magnetization using spin currents generated through spin–orbit coupling in adjacent materials. When an in-plane charge current flows through a heavy metal or quantum material with strong spin–orbit coupling, a transverse spin current is generated via the spin Hall effect (SHE). This spin current can be expressed as [21],

$$\mathbf{J}_s = \frac{\hbar}{2e}\theta_{SH}\left(\mathbf{J}_c \times \hat{\sigma}\right) \tag{1}$$

where $\mathbf{J}_s$ is the spin current density, $\mathbf{J}_c$ is the charge current density, θ_{SH} is the spin Hall angle, and $\hat{\sigma}$ represents the spin polarization direction. The converse of the SHE is the inverse spin Hall effect (ISHE), in which a spin current is converted into a transverse charge current. The spin current generated via SHE leads to spin accumulation at the interface between the high SOC layer and the ferromagnetic layer. Similarly, in systems with structural inversion asymmetry, the Rashba–Edelstein effect generates a nonequilibrium spin density given by [22]

$$\mathbf{s} \propto \hat{z} \times \mathbf{J}_c \tag{2}$$

where $\mathbf{s}$ is the induced spin accumulation and $\hat{z}$ is the interface normal direction. The accumulated spins exert a spin–orbit torque on the magnetization of the adjacent ferromagnetic layer, which can be written as [19]

$$\boldsymbol{\tau}_{\mathrm{SOT}} \propto \mathbf{m} \times (\hat{\sigma} \times \mathbf{m}) \tag{3}$$

where $\mathbf{m}$ is the magnetization unit vector. This torque enables magnetization switching without passing current directly through the magnetic tunnel junction.

Unlike STT, SOT separates the read and write paths as shown in Fig. 3, which improves device reliability, reduces switching energy, and enables faster operation. These advantages make SOT based memory devices promising candidates for next generation energy efficient and high speed spintronic memory technologies. SOT-MRAM is still in research and pre-commercial development stage, but significant progress has been made in recent years. In 2023, IMEC reported extremely scaled SOT-MRAM devices with switching energy below 100 fJ and endurance $>10^{15}$ cycles, bringing the technology closer to cache memory applications [23]. Despite their promise, MRAMs face important challenges. Scaling MRAM to very high densities remains difficult due to thermal stability requirements, write-error rates, and fabrication complexity. In addition, the cost per bit, particularly when compared to mature DRAM and flash technologies, remains a significant barrier to widespread adoption. Addressing these challenges requires improvements not only in device architecture but also in the materials used for efficient spin-current generation. Following the development of SOT-MRAM, significant efforts have therefore focused on materials capable of generating efficient spin orbit torques. Conventionally, heavy metals, in particular platinum (Pt), have been extensively employed due to their strong spin–orbit coupling and large spin Hall effect, enabling efficient spin current generation. These materials have played a crucial role in demonstrating reliable SOT based switching and device prototypes. However, heavy metals present several limitations, including relatively high power consumption, limited spin-conversion efficiency, material cost, scarcity and challenges in further scaling for next-generation technologies. To overcome these limitations, recent research has increasingly shifted toward quantum materials such as topological insulators, transition metal dichalcogenides, antiferromagnets, altermagnets, complex oxides etc. [24–38]. These materials offer fundamentally new mechanisms, including spin-momentum locking, large spin Hall angles, and unconventional magnetic ordering, enabling more efficient spin–orbit torque generation with reduced switching currents [19]. As a result, quantum materials are emerging as promising candidates for next-generation SOT-MRAM and energy efficient spintronic memory technologies.

3 Quantum Materials for Next-Generation Spintronics

The performance of spintronic devices is fundamentally determined by the efficiency with which spin information can be generated, transported, manipulated, and detected. Conventional materials, such as heavy metals (e.g., Pt, W, Ta), generate spin currents via spin–orbit interactions.

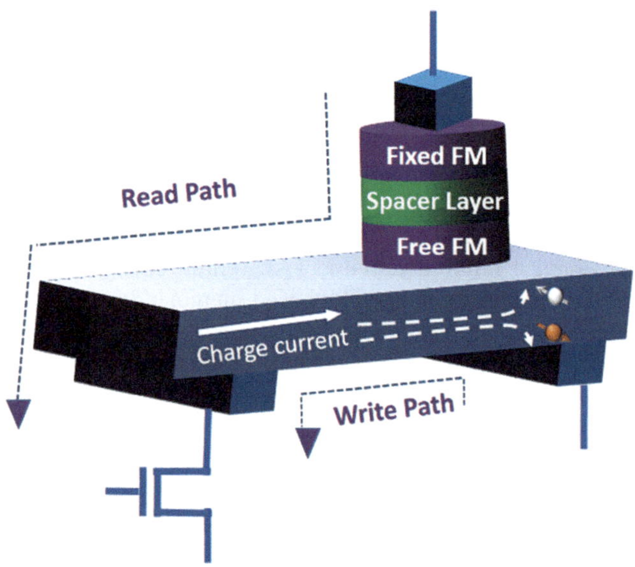

Fig. 3. Schematic illustration of a SOT-based device architecture. An in-plane charge current in the SOC layer generates a transverse spin current via spin–orbit effects, which exerts torque on the free ferromagnetic layer and induces magnetization switching. The fixed and free ferromagnetic layers, separated by a spacer layer, enable resistance-based readout through magnetoresistive effects.

However, they typically require large charge current densities, resulting in substantial Joule heating. In addition, many of these materials are scarce and not always economically viable for large-scale technological deployment. In contrast, quantum materials offer intrinsically enhanced spin-charge interconversion mechanisms arising from their unique electronic structures [24,39]. Characterized by strong electronic correlations, nontrivial band topology, and pronounced spin–orbit coupling, quantum materials provide a powerful platform for extending spintronic functionality beyond the limits of conventional systems. Features such as spin–momentum locking, reduced dimensionality, and symmetry-driven band degeneracies enable the generation of large spin currents at significantly lower charge current densities. Furthermore, quantum materials shown in Fig. 4 often host novel magnetic orders and quasiparticles, including Dirac and Weyl fermions as well as topologically protected surface states, which provide new pathways for controlling spin dynamics [24]. Collectively, these properties make quantum materials particularly well suited for addressing energy efficiency and scaling challenges in modern spintronic devices.

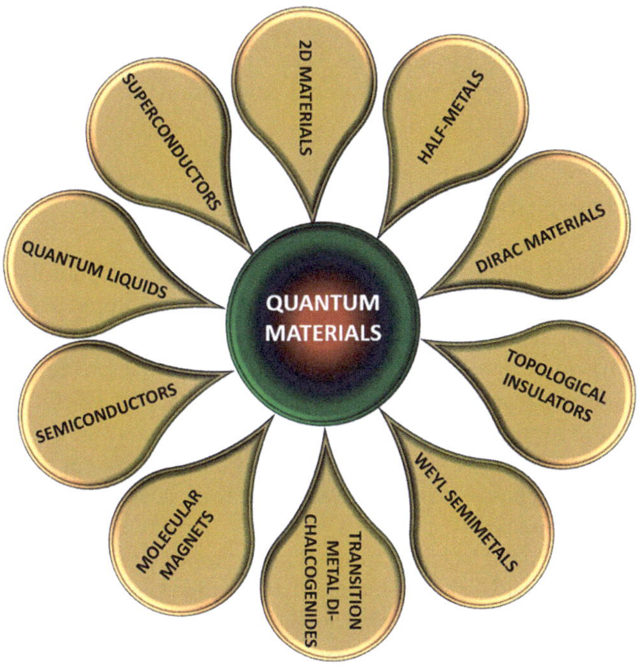

Fig. 4. Selected list of quantum materials that have drawn significant attention over last few decades.

Different classes of quantum materials offer distinct mechanisms for spin generation and manipulation. Topological insulators exhibit spin–momentum locking in their surface states, allowing highly efficient spin current generation. Transition metal dichalcogenides provide strong spin–orbit coupling and reduced dimensionality, enabling tunable spin textures in atomically thin platforms. Antiferromagnets and altermagnets have recently attracted attention due to their ultrafast spin dynamics, absence of stray magnetic fields, and robustness against external perturbations, making them promising candidates for high-density memory and logic applications. In addition, Weyl and Dirac semimetals exhibit large Berry curvature effects that can enhance spin–orbit torques and enable unconventional spin transport phenomena [40,41]. Another important advantage of quantum materials lies in their interfacial properties. Spintronic devices often rely on heterostructures where efficient spin transfer across interfaces is crucial. Quantum materials frequently exhibit high interfacial spin transparency and strong spin Hall conductivity, leading to improved spin injection efficiency [26,42]. Furthermore, their tunability through strain, gating, and layer engineering provides additional control over spin transport properties, making them attractive for scalable and CMOS-compatible device architectures [19,39]. By integrating such materials into SOT-MRAM architectures, it becomes possible to overcome current scaling and power limitations. Beyond memory technologies, quantum materials are also enabling new directions in spin-based logic, neuromorphic computing, magnonics, and quantum information technologies, expanding the role of spintronics toward multifunctional and energy-efficient computing architectures (Fig. 5).

4 Emerging Spintronic Applications Beyond Memory

While memory has been the primary driver of spintronic commercialization, spin-based logic represents a longer-term but transformative application. Spin-based logic devices aim to use spin states, rather than charge accumulation, to represent and manipulate logical information [43]. This approach promises significantly lower power consumption by reducing Joule heating and eliminating the need for continuous charge flow. Proposed spin logic architectures include all-spin logic, where spin currents transmit information between magnetic elements, and hybrid spin-CMOS schemes that integrate spintronic elements with conventional electronics. Spin transistors, in which the output current is controlled by spin orientation rather than electrostatic gating alone, exemplify this concept [44]. Although practical realization remains challenging due to issues such as spin injection efficiency and spin coherence length, progress in materials and interfaces continues to push spin-based logic closer to feasibility.

Another important application of spintronics is spin caloritronics, an emerging field that explores the interaction between heat, charge, and spin in magnetic materials, offering a promising route for harvesting waste heat from electronic devices. In modern computers and large-scale data centers, a significant portion of electrical energy is lost as heat due to Joule heating. Instead of dissipating this heat using cooling systems, spin caloritronics aims to convert it into useful electrical energy. One of the most important mechanisms enabling this conversion is the Spin Seebeck Effect (SSE), where a temperature gradient across a magnetic material generates a spin current [45]. This spin current can then be

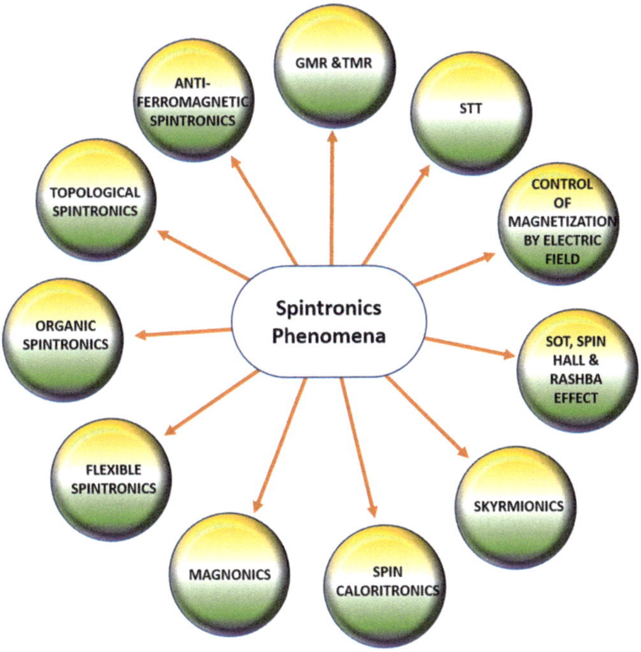

Fig. 5. An overview of different sub-topics of the emerging field of spintronics.

converted into a measurable electrical voltage by ISHE in the adjacent high SOC layer. This approach is particularly attractive because it allows energy harvesting even from small temperature differences. By integrating spin caloritronic devices into processors, memory units, and data center infrastructure, waste heat that is currently lost can be partially recovered, improving overall energy efficiency. Spintronics has also found a natural role in neuromorphic computing, which seeks to emulate brain-like information processing using hardware architectures [46,47]. Magnetic systems inherently exhibit non-volatility, history dependence, and nonlinear dynamics, features that closely resemble synaptic and neuronal behavior. Domain walls, and topological spin textures such as skyrmions have been explored as artificial synapses and neurons [48,49]. Skyrmions, in particular, are nanoscale, topologically protected magnetic structures that can be manipulated with extremely low current densities [48]. Their stability, mobility, and tunability make them attractive candidates for dense, energy-efficient neuromorphic circuits. Spintronic neuromorphic systems promise massively parallel computation with orders-of-magnitude lower energy consumption compared to conventional von Neumann architectures [50]. Spintronic devices also offer unique capabilities in radio-frequency (RF) and communication technologies [51]. Spin-torque oscillators (STOs) generate microwave signals through sustained magnetization precession driven by spin-transfer or spin–orbit torques [19].

Unlike conventional RF oscillators, STOs are nanoscale, frequency-tunable, and compatible with CMOS processes. STOs hold promise for applications in wireless communication, signal processing, and neuromorphic RF systems. Their ability to synchronize and exhibit nonlinear dynamics opens avenues for unconventional computing and communication paradigms, including oscillator-based neural networks and frequency-encoded logic [52]. At the frontier of spintronics lies its intersection with quantum information science. Electron spins in semiconductors, defects, and quantum dots serve as natural candidates for qubits due to their well-defined quantum states and potential for long coherence times [53]. Spin qubits offer compatibility with existing semiconductor fabrication technologies, providing a pathway toward scalable quantum processors. Additionally, magnonic systems and spin-wave-based qubits are being explored as alternative platforms for quantum information processing, offering long-range coherence and low dissipation [54–56]. Organic spintronics represents another important application of spintronics, where organic molecules and polymers are employed as active materials for spin transport and manipulation. In contrast to conventional inorganic materials, organic semiconductors typically exhibit weak spin–orbit coupling and reduced hyperfine interactions, which can result in longer spin relaxation times and extended spin diffusion lengths. Several organic materials, such as buckminsterfullerene, Alq_3, phthalocyanine molecules etc have been investigated for spin transport and spin valve applications [57–61]. These materials also offer additional advantages, including low-cost fabrication, chemical tunability, and compatibility with large-area processing.

Another interesting application in spintronics is the flexible spintronics that represents an emerging and interdisciplinary extension of conventional spintronic technologies, aimed at integrating spin-dependent functionalities with mechanically flexible and deformable platforms [62–64]. By combining magnetic materials, spin–orbit coupled layers, and magnetoresistive heterostructures with flexible substrates such as polymers, ultrathin foils, flexible spintronics enables devices that can bend, stretch, and conform to non-planar surfaces without compromising magnetic performance. This capability opens new application spaces beyond rigid silicon-based electronics, including wearable sensors, implantable biomedical devices, soft robotics, and human-machine interfaces [62]. From a materials perspective, flexible spintronics poses unique challenges, as magnetic anisotropy, interfacial spin propagation, and spin–charge conversion efficiency can be strongly influenced by strain, bending, and mechanical deformation. At the same time, these mechanical degrees of freedom offer new opportunities, allowing strain to be used as an active control parameter for tuning magnetic states, spin transport, and magnetization dynamics. Looking ahead, the future of flexible spintronics will depend on advances in robust magnetic thin films compatible with flexible substrates, and scalable integration with flexible electronics. As energy-efficient, non-volatile, and multifunctional devices become increasingly important for next-generation wearable and bio-integrated systems, flexible spintronics is poised to play a critical role in enabling electronics that are not only smarter and more energy efficient but also mechanically adaptive and seamlessly integrated with everyday life.

5 Summary and Outlook

Spintronics has evolved from a scientific curiosity into a versatile platform for next-generation information technologies, with quantum materials playing an increasingly central role in driving this transformation. While early spintronic devices relied heavily on conventional ferromagnets and heavy metals, the emergence of quantum materials has significantly expanded the scope and efficiency of spin-based phenomena. Materials such as topological insulators, transition metal dichalcogenides, antiferromagnets, altermagnets, and complex oxides exhibit strong spin–orbit coupling, unconventional magnetic ordering, and novel interfacial effects, enabling highly efficient spin generation, manipulation, and detection. These properties offer promising pathways toward reducing switching currents, improving scalability, and enhancing device performance beyond the limits of conventional material systems. Despite considerable progress, several challenges remain, including achieving efficient spin transport at low power, ensuring robust interfacial engineering, and integrating novel materials with CMOS-compatible fabrication processes. Device variability, material stability, and large-scale manufacturability also continue to influence the practical deployment of spintronic technologies. In this context, quantum materials provide fundamentally new opportunities to address these challenges by introducing mechanisms such as spin–momentum locking, large spin Hall effects, and exotic magnetic textures. Looking forward, the convergence of spintronics with quantum materials, advanced nanoscale engineering, and system-level design is expected to unlock new functionalities and applications. Emerging concepts such as spin–orbit torque memories, antiferromagnetic spintronics, magnonics, skyrmionics, and neuromorphic computing architectures offer promising routes toward ultrafast, energy-efficient computing technologies. As the demand for sustainable and high-performance information processing continues to grow, quantum-material-based spintronics is poised to play a transformative role in shaping future memory, logic, and computing technologies beyond conventional electronics.

Acknowledgement. We thank several of our past and present group members as well as collaborators. Further, we acknowledge funding from Department of Atomic Energy (DAE) and Department of Science and Technology (DST), Government of India.

References

1. Krūmiņš, J., Kļaviņš, M.: Energies **16**, 3612 (2023)
2. Kabeyi, M.J.B., Olanrewaju, O.A.: Front. Energy Res. **9**, 743114 (2022)
3. Saleh, H.M., Hassan, A.I.: Appl. Chem. Eng. **7**, 2084 (2024)
4. Huang, P., Copertaro, B., Zhang, X., Shen, J., Löfgren, I., Rönnelid, M., Fahlen, J., Andersson, D., Svanfeldt, M.: Appl. Energy **258**, 114109 (2020)

5. Kelechi, A.H., Alsharif, M.H., Bameyi, O.J., Ezra, P.J., Joseph, I.K., Atayero, A.-A., Geem, Z.W., Hong, J.: Symmetry **12**, 1029 (2020)
6. Gathu, S., et al.: J. Adv. Comput. Syst. **4**, 22 (2024)
7. Guo, Z., Yin, J., Bai, Y., Zhu, D., Shi, K., Wang, G., Cao, K., Zhao, W.: Proc. IEEE **109**, 1398 (2021)
8. Incorvia, J.A.C., Xiao, T.P., Zogbi, N., Naeemi, A., Adelmann, C., Catthoor, F., Tahoori, M., Casanova, F., Becherer, M., Prenat, G., et al.: Nat. Rev. Electr. Eng. **1**, 700 (2024)
9. Puebla, J., Kim, J., Kondou, K., Otani, Y.: Commun. Mater. **1**, 24 (2020)
10. Thompson, S.M.: J. Phys. D Appl. Phys. **41**, 093001 (2008)
11. Ryu, J., Lee, S., Lee, K.-J., Park, B.-G.: Adv. Mater. **32**, 1907148 (2020)
12. Pathak, S., Youm, C., Hong, J.: Sci. Rep. **10**, 2799 (2020)
13. Hoffmann, A.: IEEE Trans. Magn. **49**, 5172 (2013)
14. Wang, K., Alzate, J., Amiri, P.K.: J. Phys. D Appl. Phys. **46**, 074003 (2013)
15. Haigh, T.: IEEE Ann. Hist. Comput. **31**, 6 (2009)
16. Seagate Technology. https://www.seagate.com/in/en/products/enterprise-drives/exos/ (2025)
17. Computer History Museum. https://www.computerhistory.org/storageengine/solid-state-drive-module-demonstrated/ (1991)
18. KIOXIA Corporation: KIOXIA Announces Industry's First 245.76 TB NVMe SSD Built for Generative AI Environments. https://www.kioxia.com/en-jp/business/news/2025/20250722-1.html (2025)
19. Shao, Q., Li, P., Liu, L., Yang, H., Fukami, S., Razavi, A., Wu, H., Wang, K., Freimuth, F., Mokrousov, Y., et al.: IEEE Trans. Magn. **57**, 1 (2021)
20. Everspin Technologies, Inc.: Everspin Expands PERSYST MRAM Family with High-Reliability Devices for Aerospace, Defense, Automotive and Industrial Applications. https://investor.everspin.com/news-releases/news-release-details/everspin-expands-persyst-mram-family-high-reliability-devices (2025)
21. Saitoh, E., Ueda, M., Miyajima, H., Tatara, G.: Appl. Phys. Lett. **88**, 498 (2006)
22. Manchon, A., Koo, H.C., Nitta, J., Frolov, S.M., Duine, R.A.: Nat. Mater. **14**, 871 (2015)
23. Imec: Imec's Extremely Scaled Sot-MRAM Devices. https://www.imec-int.com/en/press/imecs-extremely-scaled-sot-mram-devices-show-record-low-switching-energy-and-virtually (2023)
24. Han, W., Otani, Y., Maekawa, S.: NPJ Quant. Mater. **3**, 27 (2018)
25. Chen, H., Yi, D.: APL Mater. **9**, 598 (2021)
26. Mishra, A., Rathore, K.S., Mahanta, S.P., Bedanta, S.: Phys. Rev. B **113**, 014407 (2026)
27. Zhang, Y., Bai, H., Dai, J., Han, L., Chen, C., Liang, S., Cao, Y., Zhang, Y., Wang, Q., Zhu, W., et al.: Nat. Commun. **16**, 5646 (2025)
28. Mishra, A., Das, P., Chhatoi, R., Dash, S., Sahoo, S., Rathore, K.S., Cha, P.-R., Lee, S.-C., Bhattacharjee, S., Bedanta, S.: Phys. Rev. Appl. **24**, 064032 (2025)
29. Mishra, A., Gupta, P., Thiruvengadam, V., Singh, B.B., Bedanta, S.: J. Alloy. Compd. **970**, 172076 (2024)
30. Bangar, H., Kumar, A., Chowdhury, N., Mudgal, R., Gupta, P., Yadav, R.S., Das, S., Muduli, P.K.: ACS Appl. Mater. Interf. **14**, 41598 (2022)
31. Paul, D., Yadav, S., Gupta, S., Patra, B., Kulkarni, N., Mondal, D., Gavankar, K., Sahu, S.K., Satpati, B., Singh, B., et al.: Adv. Sci. **12**, e02569 (2025)
32. Roy, K., Mishra, A., Nayak, S., Gupta, P., Jena, B.B., Bedanta, S.: ACS Appl. Electron. Mater. **5**, 1575 (2023)

33. Singh, B.B., Roy, K., Chelvane, J.A., Bedanta, S.: Phys. Rev. B **102**, 174444 (2020)
34. Singh, B.B., Bedanta, S.: Phys. Rev. Appl. **13**, 044020 (2020)
35. DuttaGupta, S., Kurenkov, A., Tretiakov, O.A., Krishnaswamy, G., Sala, G., Krizakova, V., Maccherozzi, F., Dhesi, S., Gambardella, P., Fukami, S., et al.: Nat. Commun. **11**, 5715 (2020)
36. Manchon, A., Železnỳ, J., Miron, I.M., Jungwirth, T., Sinova, J., Thiaville, A., Garello, K., Gambardella, P.: Rev. Mod. Phys. **91**, 035004 (2019)
37. Singh, B.B., Jena, S.K., Samanta, M., Biswas, K., Satpati, B., Bedanta, S.: Phys. Status Solidi Rapid Res. Lett. **13**, 1800492 (2019)
38. Duan, J., Hu, S., Wang, P., Zhang, D., Jiang, Y.: MetalMat **1**, e24 (2024)
39. Guo, Y., Zhang, X., Huang, Z., Chen, J., Luo, Z., Zhang, J., Li, J., Zhang, Z., Zhao, J., Han, X., et al.: NPJ Spintron. **2**, 36 (2024)
40. Ding, J., Liu, C., Kalappattil, V., Zhang, Y., Mosendz, O., Erugu, U., Yu, R., Tian, J., DeMann, A., Field, S.B., et al.: Adv. Mater. **33**, 2005909 (2021)
41. Li, P., Wu, W., Wen, Y., Zhang, C., Zhang, J., Zhang, S., Yu, Z., Yang, S.A., Manchon, A., Zhang, X.-X.: Nat. Commun. **9**, 3990 (2018)
42. Zhu, D., Wang, Y., Shi, S., Teo, K.-L., Wu, Y., Yang, H.: Appl. Phys. Lett. **118**, 598 (2021)
43. Bandyopadhyay, S., Cahay, M.: Nanotechnology **20**, 412001 (2009)
44. Datta, S.: Nat. Electr. **1**, 604 (2018)
45. Uchida, K., Takahashi, S., Harii, K., Ieda, J., Koshibae, W., Ando, K., Maekawa, S., Saitoh, E.: Nature **455**, 778 (2008)
46. Sengupta, A., Roy, K.: Appl. Phys. Express **11**, 030101 (2018)
47. Sengupta, A., Roy, K.: Appl. Phys. Rev. **4**, 1158 (2017)
48. Zhou, Y., Li, S., Liang, X., Zhou, Y.: Adv. Mater. **37**, 2312935 (2025)
49. Liu, L., Wang, D., Wang, D., Sun, Y., Lin, H., Gong, X., Zhang, Y., Tang, R., Mai, Z., Hou, Z., et al.: Nat. Commun. **15**, 4534 (2024)
50. Plummer, D.Z., D'Alessandro, E., Burrowes, A., Fleischer, J., Heard, A.M., Wu, Y.: J. Low Power Electr. Appl. **15**, 16 (2025)
51. Dieny, B., Prejbeanu, I.L., Garello, K., Gambardella, P., Freitas, P., Lehndorff, R., Raberg, W., Ebels, U., Demokritov, S.O., Akerman, J., et al.: Nat. Electr. **3**, 446 (2020)
52. Csaba, G., Porod, W.: Appl. Phys. Rev. **7**, 789–982 (2020)
53. Kloeffel, C., Loss, D.: Annu. Rev. Condens. Matter Phys. **4**, 51 (2013)
54. Flebus, B., Grundler, D., Rana, B., Otani, Y., Barsukov, I., Barman, A., Gubbiotti, G., Landeros, P., Akerman, J., Ebels, U., et al.: J. Phys.: Condens. Matter **36**, 363501 (2024)
55. Rathore, K.S., Swain, A., Lim, J., Mishra, A., Gupta, P., Yong-Heng, L., Luwen, J., Mahendiran, R., Hoffmann, A., Bedanta, S.: Phys. Rev. Appl. **25**, 024036 (2026)
56. Barman, A., Gubbiotti, G., Ladak, S., Adeyeye, A.O., Krawczyk, M., Gräfe, J., Adelmann, C., Cotofana, S., Naeemi, A., Vasyuchka, V.I., et al.: J. Phys.: Condens. Matter **33**, 413001 (2021)
57. Mallik, S., Mattauch, S., Dalai, M.K., Brückel, T., Bedanta, S.: Sci. Rep. **8**, 5515 (2018)
58. Sahoo, A., Mukhopadhyaya, A., Mahanta, S.P., Ali, M.E., Bedanta, S.: Phys. Rev. Appl. **21**, 054001 (2024)
59. Pandey, E., Sharangi, P., Sahoo, A., Mahanta, S.P., Mallik, S., Bedanta, S.: Appl. Phys. Lett. **123**, 589 (2023)
60. Pandey, E., Ojha, B., Bedanta, S.: Phys. Rev. Appl. **19**, 044013 (2023)
61. Mahanta, S.P., Sahoo, A., Mukhopadhyaya, A., Nayak, S., Hase, T.P., Ali, M.E., Atkinson, D., Bedanta, S.: ACS Appl. Nano Mater. **8**, 12630 (2025)

62. Bo, G., Xu, X., Meng, K., Wu, Y., Jiang, Y.: NPJ Spintron. **3**, 48 (2025)
63. Chen, X., Mi, W.: J. Mater. Chem. C **9**, 9400 (2021)
64. Pandey, E., Singh, B.B., Sharangi, P., Bedanta, S.: Nano Express **1**, 010037 (2020)
65. https://www.mram-info.com/stt-mram

Indigenous Material Technology for Sustainable Power Generation: Materials for FBR and AUSC Power Plants

Dipti Samantaray[1,2(✉)], B. Aashranth[1], Arvinth Davinci[1], Mantosh Mandal[1,2], N. L. Parthasarathi[1], V. Karthik[1,2], A. Nagesha[1,2], M. Vasudevan[1,2], and Anish Kumar[1,2]

[1] Indira Gandhi Centre for Atomic Research, Kalpakkam 603102, India
deepsaroj@igcar.gov.in
[2] Homi Bhabha National Institute, Mumbai 400094, India

Abstract. The simultaneous achievement of India's growth and sustainable development goals, in the form of increased power generation and decarbonization of the power sector, largely depends on the indigenous capability to manufacture and process materials. This includes structural and functional materials, both for nuclear power plants and for advanced coal-based power plants. An overview of the range of materials used in both types of power plants is provided, with emphasis paid on the current status of indigenization and the work currently being carried out in this direction. The challenges faced during indigenization at scale, and the lessons learnt from recent projects related to fast breeder reactor materials and advanced ultra-super critical power plant materials, are discussed.

Keywords: materials technology · manufacturing · indigenization · structural materials · sustainable energy

1 Introduction

In the year 2025, India's electricity consumption peaked at 241.2 GW with the per capita consumption of 1460 kWh [1]. The present installed capacity (~ 510 GW) could facilitate the availability of electricity for an average of 23.4 h per day in urban areas and 22.6 h per day in rural areas [1, 2]. These data reveal that the presently installed capacity is insufficient to ensure 24/7 electricity supply throughout the country.

In order to elevate the living standards, expand national infrastructure, and support industrial growth, it is necessary that the per capita energy consumption must increase significantly. For example, the electricity demand is forecasted to reach 708GW by 2047, if the GDP growth rate is sustained at 7%. To satisfy this requirement under a 'Net Zero' scenario, India must scale up its total installed capacity to ~ 2120 GW [3]. Simultaneously, decarbonization targets require that the energy share of fossil-fuel sources be reduced from the present ~ 50 to ~ 10%. An important contribution to the non-fossil fuel sources comes from nuclear energy, which is forecasted to grow to 100 GW.

G. B. Mohanty and V. S. Prabhudesai (Eds.): DAEC 2026, SPPHY 368, pp. 99–106, 2026.
https://doi.org/10.1007/978-981-92-2948-2_11

The success of energy expansion and sustainable power generation is significantly dependent on the structural materials used in various types of power plants. These structural materials determine the operational efficiency, longevity, reliability and capital cost of power production. It is therefore critical that all aspects of materials technology associated with these materials be indigenously developed, and self-sufficiency be achieved at every stage of the manufacturing process. In this article, the progress made towards indigenization is discussed, with major emphasis on materials technologies for nuclear fast breeder reactors and Indian advanced ultra-super critical power plants (Table 1).

Table 1. India's installed power capacity in 2025 and required installed power capacity in 2047, in Net Zero scenario, assuming 7% GDP growth

Energy sources	Installed capacity (GW) in 2025 [2]	Required installed capacity (GW) in 2047 [3]
Gas	20.71	24.96
Coal	226.23	176.49
Nuclear	8.78	100.46
Hydro	56.07	112.20
Solar	135.80	1233.59
Wind	54.51	452.39
Biomass and others	11.61	20.85
Total	513.72	2120.94

2 Evaluating Material Demands of Power Plants

The choice of structural materials primarily depends on three major factors: (a) the type of fuel, i.e., the heat source, (b) the working fluid, and (c) the operational parameters. These factors are discussed for nuclear fast breeder reactors (FBRs) and coal-based advanced ultra-super critical power plants (AUSCPP) in Table 2. It may be inferred that some property requirements are common to both types of power plants, whereas certain requirements are unique to each. For example, structural materials in both types of power plants require to be creep resistant. On the other hand, resistance to radiation damage is critical only for the FBR structural materials, whereas resistance to fire-side corrosion is essential only in AUSCPP structural materials.

Table 2. Major features and operational parameters of AUSC and FBR-based power plants

Feature	Operating parameters		Properties needed in structural material	
	AUSCPP	FBR	AUSCPP	FBR
Fuel	Coal	U-Pu based mixed oxide/mixed carbide/metallic fuel	–	Resistance to radiation damage
Primary fluid (medium of heat transport from fuel to working fluid)	Flue gas	Liquid metal (typically liquid Na)	Resistance to fire-side corrosion	Compatibility with liquid metal
Operating temperature for primary fluid	120–1500 °C [4]	485–546 °C	Creep resistance	Creep resistance
Working fluid	Water/steam	Water/steam	Resistance to water and steam side corrosion	Resistance to water and steam side corrosion
Operating parameters for the working fluid	700–720 °C, 31–35 MPa [5]	460–520 °C, 12.5–17 MPa [6, 7]	Resistance to creep and thermal fatigue	Resistance to creep and thermal fatigue

2.1 Structural Materials for FBR-Based Power Plants

In FBRs, the heat needed for the power generation is largely produced from the fission of Pu^{239} and U^{235} isotopes. These fissile isotopes are used as FBR fuels in oxide, carbide or metallic forms. The form of the fissile fuel directly influences the FBR's operating temperature, breeding ratio (the number of fissile atoms produced from each fissile atom consumed) and the neutron spectrum (the magnitude and spread of neutron energies). The use of oxide fuel permits higher operating temperature; this allows for higher thermal efficiency, but also constrains the breeding potential of the reactor. On the other hand, the metal fuel enhances breeding performance yet imposes a limit on operating temperature due to its low melting point. Since thermal efficiency and breeding ratio are equally vital to the broader objective of increasing nuclear power share in India's energy mix, different FBRs are designed to use different fuels. To fully leverage the capabilities of various FBR cores, the structural materials and their associated technologies must evolve in tandem with the selection of nuclear fuel.

To facilitate a structured overview, we have classified the FBR structural materials into three broad categories: core structural materials, near-core structural materials and SG material. The examples of the 'core structural materials' are the materials used to make the fuel cladding, the spacer wires wrapped around each fuel pin, and the wrappers which encapsulate groups of fuel pins for efficient handling. These components must withstand elevated temperature and intense neutron irradiation; however, these are retained in the core only for short durations. The 'near-core structural materials' are the materials used for the construction of FBR components such as the intermediate heat exchanger (IHX) and the main vessel/safety vessel. These components sustain lower levels of irradiation relative to the core structural materials. While some of these components are replaceable, others are designed for life time.

2.1.1 Core Structural Materials

In the Indian context, the operation of FBRs to generate power was demonstrated with U-Pu carbide fuel based test reactor, namely the Fast Breeder Test Reactor (FBTR). The core of FBTR operates in the temperature range of 380–510 °C, with the maximum clad temperature of 600 °C [6]. For this operating temperature, SS 316M is used as clad, spacer wire and wrapper material. FBTR is also used as a test bed to study the performance of candidate fuel and structural materials for in-core applications in future FBRs. The data generated from the FBTR has validated the choice of the materials for the design of a 500 MWe FBR based power plant, namely the Prototype Fast Breeder Reactor (PFBR). The PFBR uses oxide fuel and is designed to operate in temperature range 400–546 °C. Given that this reactor operates at a slightly higher temperature than the FBTR and clad hot spot temperature here is expected to be 720 °C, alloy D9 was chosen as the clad, spacer wire and wrapper material. Alloy D9 with an optimum Ti/C ratio provides greater resistance to void swelling (increase in volume due to structural damage caused by neutron bombardment) than SS 316M in the operating temperature range of the PFBR, and also has superior creep resistance.

Irradiation data further suggests that enhancing Si and P content in D9 can further improve its void swelling resistance. These data led to development of Indian Fast reactor clad material (IFAC-1). IFAC-1 exhibits better swelling resistance compared to alloy D9 while maintaining comparable high temperature properties. While the irradiation damage tolerance capability of IFAC-1 is superior to the other austenitic stainless steel candidate materials, it remains insufficient to achieve the maximum attainable burn-up of the oxide fuel. Since the swelling susceptibility of austenitic stainless steels is dictated by the FCC structure, further improvement through alloying additions may not be significant, unless the crystal structure is fundamentally altered.

On the other hand, the ferritic/martensitic steels, which have a BCC structure, exhibit inferior creep properties compared to the austenitic stainless steels. However, these steels show remarkably superior void swelling resistance. Therefore, efforts are being made to enhance the creep properties of the ferritic/martensitic steels through dispersion of fine Y_2O_3 particles in the matrix. This class of steels is known as Oxide Dispersion Strengthened (ODS) steel. As part of ODS steel development, steels with 9–18 wt.% Cr are being studied. The initial studies suggest that 9% Cr containing ODS steel dissolves in boiling nitric acid during aqueous reprocessing of the spent nuclear fuel, whereas 18% Cr containing ODS steel exhibits poor manufacturability. Therefore, both low-Cr (~9% Cr) and high-Cr (~18% Cr) ODS steels are deemed unsuitable for the intended application. At present, the properties of various mid-Cr ODS steels are being evaluated for FBR applications.

For the metallic fuel-based reactors, which will operate in the temperature range of 500 °C with expected peak clad temperature of 650 °C, ferritic/martensitic steels containing 9% Cr are considered suitable in view of their excellent resistance to void swelling. However, ferritic/martensitic steels exhibit a high susceptibility to fuel-clad chemical interaction (FCCI) with metallic fuel. To prevent FCCI and to effectively transfer the heat from this low melting point fuel to the coolant, two types of pin designs are being considered. These are the sodium-bonded fuel pins (where liquid sodium is introduced between the U-Pu-Zr fuel and the clad) and the mechanically-bonded fuel pins (where the U-Pu fuel is directly in contact with the clad). For the sodium-bonded fuel pins, modified 9Cr-1Mo steel (Grade 91) and Grade 92 steels are the candidate clad materials. For the mechanically-bonded fuel pins, bimetallic tubes comprising Grade 91 tube with an internal Zircaloy 4 liner are being developed.

In addition to the fuel sub-assemblies, the control and safety sub-assemblies are also essential constituents of FBR cores. Both PFBR and FBTR use B_4C with various levels of B^{10} enrichment. The exact isotopic composition of B_4C pellets and their desired mechanical properties are optimized based on the specific core configuration and design.

2.1.2 Near Core Structural Materials

In the FBTR, SS 316 is used for grid plate, reactor vessel and piping [6]. In contrast, the PFBR comprises grid plate, IHX and the main vessel made using SS 316 LN (0.06–0.08 wt.% nitrogen). This choice was primarily made to avoid sensitization during welding. The main vessel of PFBR is contained in the safety vessel [8], which is made using SS 304LN, as this is not exposed to high temperature. Recent research work reveals a positive correlation between nitrogen content enhancement and long term creep strength in SS 316L(N). The suitability of nitrogen enhanced SS 316L(N) for the structural applications in future reactors is currently under evaluation.

The IHX facilitates the exchange of heat between the primary and secondary sodium circuits. In PFBR, multiple IHXs are suspended from the roof slab into the main vessel. It is imperative to shield these IHXs from the core to minimize the activation of the sodium in the secondary circuit. This is achieved through a dual-shielding approach, incorporating both core shielding and dedicated IHX-shielding. While core shielding is realized by incorporating ferroboron subassemblies around the active core [9], SS 304B (borated stainless steel) is used for the IHX shielding. Sometimes, local shielding of major components is necessary for reducing the irradiation damage. For example, WC has been recently introduced as a shield to curtail the radiation damage to the FBTR grid plate. This arrangement has been made to enhance the service life of the grid plate, thereby extending the service life of the reactor.

2.1.3 Steam Generator (SG) Materials

Steam generator (SG) of FBR is a shell-and-tube type heat exchanger which contains liquid sodium on the shell side and steam-water on the tube side. The violent chemical reaction between sodium and water demands stringent material selection for the barrier between the fluids, namely the SG tubes. The selection criteria for the SG material have been listed in the reference [8], and is summarized here. 2.25 Cr-1Mo steel was used to make the SG in the FBTR [6]. The higher steam temperatures of PFBR necessitate use of material with enhanced creep properties. Therefore, Grade 91 was selected for the application. While extensive experience is available for this material, further improvements are being sought to reduce the occurrence of Type-IV cracking in Grade 91 steel welds. Recent observations suggest that the controlled addition of boron (130–140 ppm) and control of nitrogen (80–100 ppm) may serve this purpose. Steels with these chemical compositions are being developed for future FBRs.

2.2 Structural Materials for Fossil Fuel-Based Power Plants

India has launched a major project to build a high efficiency fossil fuel-based power plant using Advanced Ultra Supercritical (AUSC) technology. To achieve thermal efficiency above 45%, a minimum steam pressure of 31 MPa and temperature 720 °C is targeted at the turbine inlet [5]. In this high temperature regime, the conventional structural materials such as 9–12% Cr ferritic/martensitic steels (typically designed for operation upto 600 °C and 31 MPa) would not be suitable. In view of this, SS 304HCu and alloy 617M (a Ni-base super alloy) were indigenously developed for super heater tubes, reheater tubes and turbine applications of the AUSC power plants. To facilitate rise in operating temperature above 720 °C and pressure at 35 MPa, another Ni-base super alloy called Indian Advanced High Temperature Alloy (IHTA) was indigenously developed. IHTA is also being evaluated as a candidate material for high temperature gas-cooled reactor (HTGR) operating with supercritical carbon dioxide (sCO_2) Brayton cycle.

2.3 The Role of Manufacturing in Alloy Development

The development of new alloys is fundamentally dependent on concurrent advancement in manufacturing technology. Our recent experience with 18Cr ODS steel development reaffirms that manufacturability often dictates the practical viability of a new alloy. For the established alloys, where manufacturability is well assured, the primary difficulty lies in optimizing the process parameters to achieve the desired properties. This issue was encountered during manufacturing of the Large Rotatable Plug /Small Rotatable Plug bearing balls for PFBR. The AISI 52100 steel is a standard alloy used for the manufacturing of bearing balls. However, the properties required for the FBR application surpassed the requirements typically established for commercial bearing balls. To meet the specified properties, an innovative processing route was established [10].

Property enhancement is frequently realized through the modification of the base composition as demonstrated in development of IFAC-I, P91B and high nitrogen variants of SS 316LN. These compositional modifications often change the workability of the material. For instance, nitrogen enhancement increases propensity for shear band formation during deformation [11]. These bands introduce inhomogeneity in the material during production. Retention of these localized bands compromises the structural integrity, thereby reducing the operational lifespan of the component. In such circumstances, rigorous process optimization becomes imperative.

Manufacturing becomes increasingly challenging when service temperature of an alloy approaches the hot working regime. These challenges were extensively encountered during manufacturing of Ni-base superalloys 617M and IHTA [12]. The manufacturing difficulty also vary with the geometry of the product. While creep resistance is essential for the tubular product, the microstructural homogeneity becomes prerequisite for the acceptance of forgings of the same superalloy. Such objectives dictate not only the meticulous optimization of thermo-mechanical and thermal processes involved in manufacturing, but also the sequence of operations in multi-stage processing chains.

The major challenge in process design is implementing the lab-scale results at industrial scale. The industrial operations are often governed by many constraints including equipment capacity, foot print of heavy machineries and their spatial positioning, and degree of automation. Therefore, effective optimization depends on designing experiments that respect the boundaries of full-scale production. Moreover, the success of a manufacturing technology necessitates a unified feedback loop between the laboratory and shop floor.

Apart from shaping the components, power plants require the establishment of robust joining and machining technologies. While precision machining is essential for effective load and motion transfer, an optimized welding process is paramount for ensuring the structural integrity and operational longevity. Besides these, development of various coating and hardfacing technologies are also important to meet the functional requirements of the power plant components.

3 Summary

To fulfill the national energy targets within a Net Zero framework, the power generation infrastructure needs to be expanded significantly. The self-sufficiency in material development and manufacturing must expand in tandem with the power infrastructure trajectory. Realization of advanced materials and associated technology necessitates a robust synergy between R&D laboratories and industry.

Acknowledgment. Authors express their gratitude to current and former colleagues from IGCAR, NFC, MIDHANI, HWB, ARCI and other institutions. Their diverse contributions have been instrumental in developing the foundational knowledge presented here.

References

1. https://www.pib.gov.in/PressReleasePage.aspx?PRID=2215187&reg=3&lang=1
2. https://iced.niti.gov.in/energy/electricity/generation/capacity
3. India energy security scenarios (IESS) 2047, Version 3.0. a Niti Ayog Report (2023)
4. Sun, E., Xu, J., Li, M., Liu, G., Zhu, B.: Connected-top-bottom-cycle to cascade utilize flue gas heat for supercritical carbon dioxide coal fired power plant. Energy Convers. Manage. **172**, 138–154 (2018)
5. Singh, K.: Materials for Fossil-fuel Fired Boilers (FFFBs). IIM Metal News **24**(4), 6–22 (2021)

6. Fast Breeder Test Reactor: An exciting journey through a new technology, SIRD, IGCAR, Kalpakkam, (2010)
7. Suresh Kumar, V.A., Noushad, I.B., Rajan, K.K.: Steam generator test facility—a test bed for steam generators of Indian sodium cooled fast breeder reactor. Nucl. Eng. Des. **248**, 169–177 (2012)
8. Mannan, S.C.: Chetal, Baldev Raj and Bhoje, selection of materials for prototype fast breeder reactor. Trans. Indian Inst. Met. **56**(2), 155–178 (2003)
9. Jayaraj, V.V., et al.: Irradiation behaviour of ferroboron—an alternate in-core shielding material for sodium-cooled fast reactors. Nucl. Eng. Des. **377**, 111126 (2021)
10. Aashranth, B., et al.,: Hardening heat treatment for AISI 52100 steel bearing balls used in rotatable plugs of PFBR, PFBR/31320/DN/1016/RevA (2024)
11. Kumar, S., Aashranth, B., Davinci, M.A., Samantaray, D., Borah, U., Bhaduri, A.K.: Formation of deformation bands in hot-worked SS 316LN. JOM **72**(3), 1395–1405 (2020)
12. Aashranth, B., Arvinth Davinci, M., Mandal, M., Samantaray, D., Karthik, V.: Hot deformation of repeat-solution annealed IHTA billets from heat nos. S4158/S4159GCAR AUSC TD-50/00 (2024)

Next-Generation Soft Semiconductor Materials and Devices: From Molecular Design to Scalable Optoelectronics

Shubham Gupta, Brijesh K. Patel, Amogh K. Ravi, Shreya Pandit, and Pabitra K. Nayak(✉)

Tata Institute of Fundamental Research, 36/P, Gopanpally Village, Hyderabad 500046, India
pabitra.nayak@tifrh.res.in

Abstract. Soft semiconductors (SoS), including organic semiconductors, halide perovskites, conjugated polymers, molecular crystals, and van der Waals heterostructures, constitute a transformative class of materials for next-generation optoelectronic and energy conversion technologies. Despite their promise, limitations in charge-carrier generation, transport, and long-term stability have hindered their widespread deployment. In a series of works, we identified predictive materials design strategies that integrate electronic doping, molecular-level control, interfacial charge transfer, crystallisation, and structure–function relationships. Central to our approach is the development of new paradigms for electronic doping of the emerging class of semiconductors, including adduct-based systems, in situ regenerative adduct-assisted (IRAA) doping, split-functional dopants, and shape-changing sublimable dopants, which together enable near-unity free-carrier generation in organic semiconductors. Complementary advances in electro-synthetic fabrication (EPIC-Fab) and the scalable design of conducting and semiconducting polymers further enable stable, scalable thin-film fabrication. Together, these contributions open many pathways towards efficient, stable, and ambient-processable optoelectronic devices based on the emerging class of semiconductors.

Keywords: Predictive materials design strategies · electronic doping · semiconductors · electro-synthetic fabrication · semiconducting and conducting polymer

1 Introduction

Soft semiconductors have emerged as a compelling alternative to conventional inorganic semiconductors owing to their tunable electronic properties, low-temperature processing capabilities, and compatibility with large-area and flexible device architectures [1]. This diverse class encompasses small-molecule organic semiconductors, conjugated polymers, hybrid halide perovskites, and van der Waals heterostructures. Nonetheless, their broader technological adoption is constrained by fundamental challenges, including inefficient electronic doping, strong Coulombic interactions arising from low dielectric constants, ion migration, and susceptibility to environmental instability. Our research

G. B. Mohanty and V. S. Prabhudesai (Eds.): DAEC 2026, SPPHY 368, pp. 107–111, 2026.
https://doi.org/10.1007/978-981-92-2948-2_12

addresses these limitations through a molecular-to-device design philosophy that establishes direct, predictive links between chemical structure, intermolecular interactions, and macroscopic device performance. By integrating advances in doping chemistry, interfacial charge transfer, crystallisation dynamics, and electronic structure, we develop a cohesive approach to controlling free charge-carrier generation, transport, and stability in soft semiconductor systems.

2 Materials and Methods Design

A central contribution of this work is the development of design strategies that link molecular-level dopant engineering to device-level functionality. This approach integrates control over charge-transfer energetics and charge-carrier and counter-ion interactions to regulate charge-transport properties. A key principle emerging from this work is the decoupling of charge transfer, counter-ion stabilisation, and processability. This decoupling enables independent optimisation of electronic doping efficiency, thermal and environmental stability, and compatibility with diverse fabrication methods.

By systematically linking molecular design to opto-electronic properties, these strategies and materials enable precise control over free-carrier generation, minimisation of energetic disorder, and enhance operational stability. The resulting materials exhibit improved conductivity and reproducibility under ambient processing and operating conditions, providing a pathway toward scalable optoelectronic technologies.

3 Highly Efficient Free-Carrier Generation in Organic Semiconductor

A major breakthrough in our work is the near-complete generation of free carriers in organic semiconductors through the development of new classes of p-type dopants. Traditional doping approaches in organic systems often suffer from incomplete ionisation, leading to residual neutral species and reduced efficiency, largely due to the low dielectric constants of organic materials (Fig. 1) [2, 3]. To address this limitation, we introduced adduct-based doping strategies that enable efficient and clean charge transfer without requiring complex synthesis or processing conditions. These dopants are compatible with both solution and vapor-phase deposition techniques and can be applied across a wide range of organic semiconductor systems, resulting in significant improvements in electrical conductivity and device performance [4].

Building on this foundation, we developed the in-situ regenerative adduct-assisted (IRAA) doping approach, which represents a fundamentally new paradigm in electronic doping. In this method, the dopant is continuously regenerated via coupled chemical reactions during doping, eliminating the need for external additives or activation steps [5]. The process is rapid, occurs under ambient conditions, and achieves near-unity doping efficiency. The absence of residual dopants and byproducts significantly enhances device stability, while the modular nature of the system allows precise tuning of ionic interactions and compatibility with different semiconductor matrices.

Further advances were achieved by introducing split-functional dopants, in which the roles of charge transfer and charge compensation are separated within a single dopant

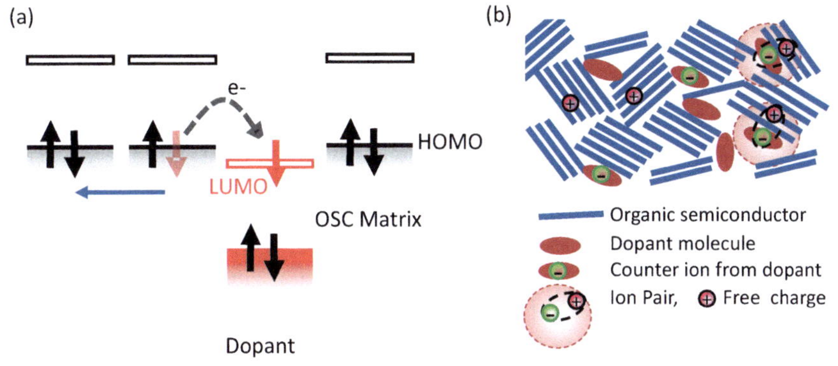

Fig. 1. Charge transfer process in a typical organic semiconductor and molecular dopant. HOMO = Highest molecular orbital, LUMO = Lowest unoccupied molecular orbital. (b) Schematic for doped organic semiconductors. The formation of ion-pairs due to Coulomb attraction leads to inefficient charge generation

architecture. This design enables independent control over redox properties, solubility, and ionic stabilization, resulting in improved doping efficiency and suppression of ion migration. The resulting films exhibit enhanced thermal stability and maintain their structural and electronic integrity over extended operational periods. These dopants can be readily integrated into solution-processed devices without the need for additional additives or complex fabrication steps.

To extend these concepts to vacuum-deposited systems, we also developed sublimable dopants that retain the advantages of modular design while enabling incorporation into multilayer device architectures. Collectively, these innovations define a scalable and versatile doping platform that significantly advances the performance and reliability of organic semiconductor devices.

4 Control of Ionic Interactions and Environmental Stability

In addition to improving doping efficiency, controlling ionic interactions and mitigating environmental degradation is critical for the stability of soft semiconductors. Oxygen-induced trap formation is a major source of performance degradation in devices based on these materials [6]. To mitigate these effects, we introduced bi-functional organic salt additives that simultaneously scavenge superoxide species and provide benign ionic substitution [7].

This dual functionality suppresses trap formation, thereby enhancing mobility and conductivity under ambient conditions. Importantly, this strategy enables device fabrication and operation without controlled environments, improving the practicality and scalability of soft semiconductor technologies.

At a fundamental level, our work establishes direct correlations between dopant molecular structure, ionic interactions, and charge-transport dynamics. By linking these factors to carrier delocalisation and relaxation processes, we provide a mechanistic basis for rational dopant design in disordered systems.

5 Electro-Synthetic Fabrication of Perovskite Films

A significant advance in scalable materials processing is the development of electro-synthetic perovskite in-situ crystallization and fabrication (EPIC-Fab) [8]. This method integrates electro-synthesis, crystallization, and film deposition into a single aqueous process, enabling the formation of highly crystalline and conformal films. This approach overcomes the intrinsic instability of halide perovskites in water while providing a scalable route for fabricating vacancy-ordered double perovskite materials. The resulting films are well-suited for photoelectrocatalytic applications and exhibit improved structural and electronic properties. The inherent light assisted repair mechanism keeps the material stable for hundreds of hours of continuous operation.

6 Scalable N-Type Semiconducting and Conducting Polymer

In parallel, we developed a single-step synthetic strategy for producing highly conductive n-type polymer semiconductors. In this approach, polymerization and doping occur simultaneously in a single reaction system, where the byproducts act as in situ dopants. This eliminates the need for external dopants, purification steps, and multistage processing. The resulting materials exhibit high electrical conductivity, optical transparency in the visible range, and intrinsic n-type transport behavior. This integrated synthesis approach significantly simplifies fabrication and enhances scalability, making it attractive for large-area electronic applications.

7 Device Integration and Applications

The materials and design strategies developed in this work have been successfully implemented across a range of optoelectronic devices, including organic thin-film transistors, organic light-emitting diodes, halide perovskite solar cells, and photoelectrochemical systems. These devices demonstrate improved efficiency and enhanced operational stability. With the new electronic dopant systems, we achieved solar cell efficiencies of ~ 25% and stability over 2000 h. under continuous operation.

A key advantage of our materials and methods is their compatibility with ambient processing and large-area fabrication, which is essential for low-cost, large-scale device production. By addressing both fundamental and technological challenges, this work bridges the gap between laboratory-scale discoveries and the real-world utilization of soft-semiconductor-based devices.

8 Conclusions and Outlook

This body of work establishes a comprehensive and predictive approach to the design of soft semiconductor materials and devices. By integrating innovations in doping chemistry, interfacial engineering, crystallization control, and scalable synthesis, we demonstrate a unified pathway for achieving efficient, stable, and scalable optoelectronic systems.

The ability to achieve near-unity free-carrier generation, control ionic interactions, and enable ambient-compatible processing marks a significant step forward in the field. These advances not only improve device performance but also open new opportunities for the development of next-generation solar energy conversion and other optoelectronic technologies. Future research will focus on extending these design principles to other emerging material systems, integrating data-driven approaches for materials discovery, and developing fully scalable and flexible device architectures. The new strategies and material systems presented here provide a strong foundation for advancing soft semiconductor technologies towards widespread applications.

Acknowledgments. S.G., B. K. P., A.K.R, S.P. and P.K.N. acknowledge support from the Department of Atomic Energy (DAE), India, under project RTI 4007.

References

1. Kirchartz, T., et al.: The state of the art in photovoltaic materials and device research. Nat. Rev. Mater. **10**, 335–354 (2025)
2. Scaccabarozzi, A.D., et al.: Doping approaches for organic semiconductors. Chem. Rev. **122**, 4420–4492 (2022)
3. Euvrard, J., et al.: The formation of polymer-dopant aggregates as a possible origin of limited doping efficiency at high dopant concentration. Org. Electron. **53**, 135–140 (2018)
4. Sakai, N., et al.: Adduct-based p-doping of organic semiconductors. Nat. Mater. **20**, 1248–1254 (2021)
5. Patel, B.K., et al.: In-Situ Regenerative Adduct Assisted P-Type Dopingof Organic Semiconductor Adv. Mater. 38 e73351 (2026). https://advanced.onlinelibrary.wiley.com/doi/10.1002/adma.73351
6. Huang, Y., et al.: Unraveling the crucial role of trace oxygen in organic semiconductors. Nat. Commun. **15**(1), 1–9 (2024)
7. Patel, B.K., et al.: Bi-functional Organic Salts for Ambient Processing of Organic Semiconductors (2026) https://doi.org/10.26434/chemrxiv.10001537/v1
8. Ravi, A.K., et al.: In-situ Electrosynthetic Fabrication of Highly Crystalline, Aqua-Compatible Palladium Halide Perovskite Films for Photoelectrocatalysis (2026) https://doi.org/10.21203/rs.3.rs-9106903/v1

Development of a Quasi Planar High Purity Ge Detector

Shreyas Pitale, Sanket Munishwar, Manoranjan Ghosh, S. G. Singh, G. D. Patra, Pradhumn Dixit, A. K. Singh, M. Sonawane, Madhu Kashyap, and Shashwati Sen(✉)

Technical Physics Division, Bhabha Atomic Research Centre (BARC), Trombay, Mumbai 400085, India
shash@barc.gov.in

Abstract. A High Purity Germanium detector in point contact quasi planar geometry is developed for gamma ray spectroscopy. The detector is fabricated using a high pure Ge crystal procured from UMICORE. The crystal is made into a diode by evaporating Li to make the n-type contact and by Boron implantation to form the p-type contacts. Further the bare surface of the crystal is passivated to achieve a leakage current of <100 pA under a reverse bias of 1500 V. The diode is mounted on a dipstick cryostat and tested at 77 K for detection of gamma rays. The energy resolution of the fabricated detector is 0.39% at 662 keV.

Keywords: Radiation · Detector · Germanium · Crystal · Diode · Passivation

1 Introduction

For ages the radiation detectors play an important role in the control and safety of nuclear reactors. However, with the increase in application of nuclear radiation in industries and healthcare, the importance of radiation detectors has also increased manifold. Among the available radiation detectors, single crystal based detectors have special importance because of their high efficiency, sensitivity, versatility and ability to identify and quantify radioactive materials. Among them High Purity Germanium (HPGe) detector have its own importance due to its high-resolution in gamma-ray spectroscopy [1].

Germanium as a semiconductor material is unique in itself because of its very low bandgap and high hole and electron mobility. The low bandgap and high mobility ensures the creation and collection of large number of electron-hole pairs (e-h) pairs after interaction with any ionizing radiation. This leads to formation a large electric signal which makes it a very sensitive detector. On the other hand the low band gap also contributes towards large noise generated due to thermally generated e-h pairs. To overcome this drawback a HPGe detector is operated at 77 K. The other challenging aspect of HPGe is the purity of the Germanium crystal which is necessary for the formation of depletion region for detection of gamma rays. A Ge crystal of purity of 99.99999999999% (13 N) is used for the fabrication of a detector. This purity pertains to active carrier concentration of ~$10^{10}/cm^3$ at 77 K in the semiconductor crystal. Additionally, for the detection of incident gamma radiation a P-I-N diode needs to be fabricated from this crystal by

G. B. Mohanty and V. S. Prabhudesai (Eds.): DAEC 2026, SPPHY 368, pp. 112–116, 2026.
https://doi.org/10.1007/978-981-92-2948-2_13

making P-type and N-type layers through deposition of Boron and Lithium coatings in the crystal. All these steps make the HPGe detector fabrication a challenging technology which is closely guarded by just two-three commercial companies all over the world.

In the endeavor to indigenously develop HPGe detectors for use in the country, a pathway was developed in which commercially available high pure Ge single crystal was used for the development of the full detector chain. A planar detector of 5 cc has been demonstrated earlier where the Ge single crystal was processed to make the Boron and Li layers [2]. Further the detector thus prepared was housed in a dipstick cryostat and connected to the required electronic chain and tested. In continuation with this research work we report the development of a HPGe detector of 50 cc active volume in point contact geometry.

2 Methods

For the fabrication of 50 cc volume HPGe diode, a high purity P type Ge crystal was procured from UMICORE, Belgium (Fig. 1(a)). As per the data sheet at 80 K, the carrier concentration of the crystal was ~1–4 × $10^{10}/cm^3$, Hall mobility ~35000 cm^2/Vs and dislocation density 3000–4000/cm^2. The crystal was cut to a diameter of 50 mm and length 25 mm with grove as shown in Fig. 1(b, c).

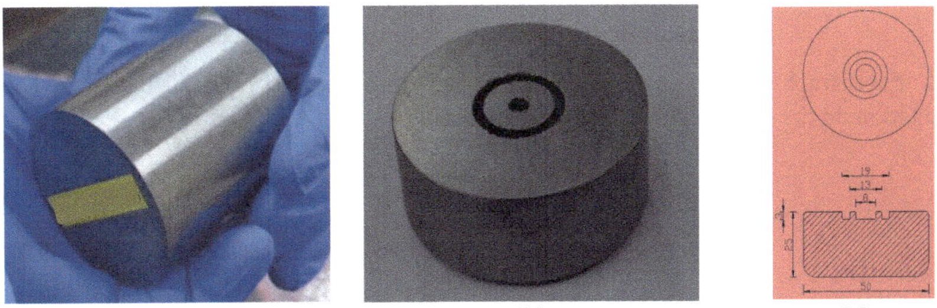

(a) (b) (c)

Fig. 1. (a) HPGe single crystal (P type) from UMICORE of 2 in. φ × 2 in. length. (b) HPGe single crystal after processing in point contact geometry. (c) Schematics of one of the many geometries tested in quasi-planar point contact geometry.

For the fabrication of the PIN diode a two step process was deployed. First the N type layer was fabricated using thermal diffusion of Li. In case of the quasi planar geometry the N type contact (rectifying contact) needs to be formed on curved surface (wrap around) and the back surface of the crystal. For this a custom made thermal evaporation system was designed with multiple electrodes to deposit Li film on one face and the curved surface simultaneously. After the deposition process the crystal was annealed in-situ for diffusion of the Li inside the Ge crystal and form a N^+ layer of thickness 500 μm.

The P type contact was formed with Boron at the centre on the top surface by Ion Implantation technique. A Danfysik (1080–30) ion implantation system was used for the purpose. The implantation was performed with ions energy of 25 keV, with dose of

10^{14} ions/cm [2]. During this process proper masking and other precautions were taken to safeguard the already deposited Li layer.

After the formation of the P and N type layer chemical passivation of grooved Surface was carried out by chemical etching and metallization was carried out by thermal deposition of Au or Al. A fully fabricated detector crystal is shown in Fig. 2.

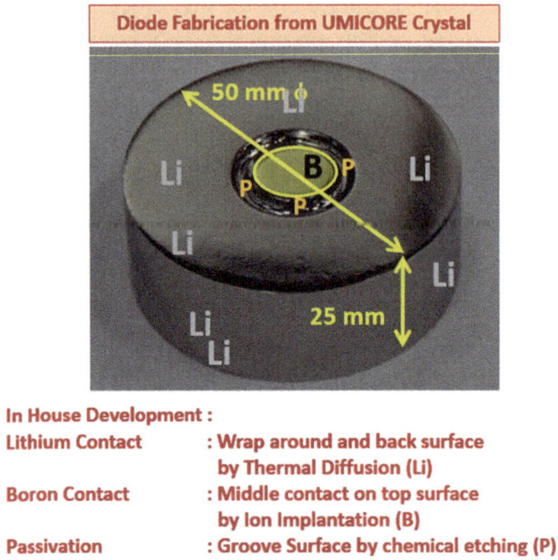

Fig. 2. Photograph of a 50 cc HPGe diode made by Li deposition on all the surface, Boron implantation on the top and chemical passivation of grove surface

3 Results

Before testing the diode in detector geometry the dark current and the breakdown voltage of the fabricated diode was tested in a low temperature IV measurement setup as shown in Fig. 3. The geometry optimized setup consists of a SS chamber to house the 2-in. diameter crystal and connected to a vacuum pump with electrical feedthrough to measure the IV characteristics. The fabricated HPGe diode is cooled using Liquid Nitrogen bath which is located at the top of the SS chamber. The fabricated diode was reversed biased to 1500 Volts and the leakage current was measured to be 80 pA showing the high quality of the formed diode and the efficacy of our procedure for the detector fabrication.

For the detector testing the developed diode was integrated with dipstick cryostat and low noise electronics as shown in Fig. 4 [2]. A custom made sample holder was designed and fabricated to hold the 2-in. diameter crystal with appropriate electrodes for biasing and signal collection. The assembled detector was tested for gamma spectroscopy for ^{137}Cs and ^{60}Co sources. The energy resolution achieved at 662 keV was 0.39% and at 1332 keV was 0.3%. Further testing is in progress to understand the effect of electronic noise and other factors on the energy resolution of the fabricated HPGe detector and further enhance it.

Fig. 3. Photograph of the setup to test the IV characteristics of the developed HPGe diode at 77 K. Magnified Image of the HV power supply showing 1500 V applied bias to the crystal

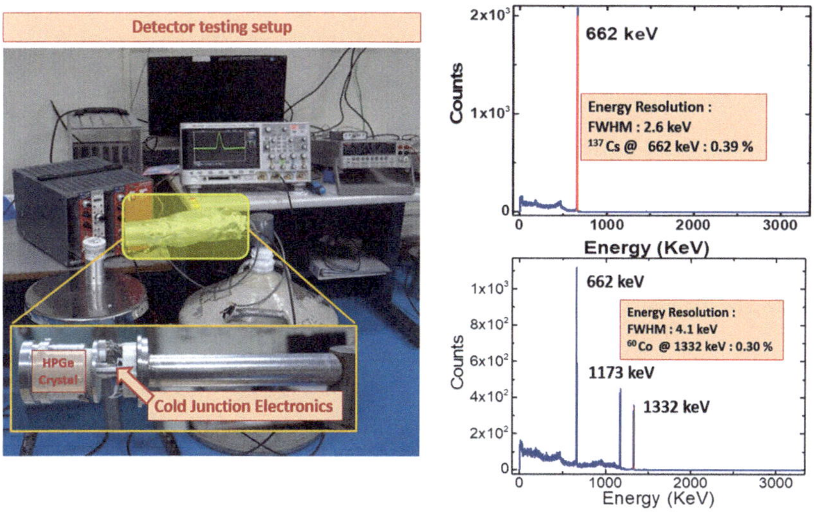

Fig. 4. (Left) Photograph of testing setup of indigenously developed 50 cc HPGe detector comprising of the HPGe diode, cold junction electronics, cryostat. (Right) the Gamma spectrum of ^{137}Cs source (Top) with FWHM 2.6 keV and energy resolution 0.39% and ^{60}Co (Bottom) with FWHM of 4.1 keV and energy resolution of 0.3%.

4 Conclusion

In summary a process was developed to fabricate a HPGe diode in quasi planar point contact geometry using a procured Ge crystal. Indigenous deposition systems were tailor made for Li and Boron deposition. The In-house fabricated detector has an active volume of 50 cc and energy resolution of ~0.39% at 662 keV.

References

1. Vetter, K.: Annu. Rev. Nucl. Part. Sci. **57**, 363–404 (2007)
2. Pitale, S. et al.: Status of indigenous development of HPGe detectors in BARC. BARC Newslett. **25**, 25–34 (2023)

Biology, Environment, and Healthcare

Metabolome Meets Microbiome: Systems Chemistry of Health and Disease

Palok Aich[1,2](✉)

[1] School of Biological Sciences, National Institute of Science Education and Research, Bhubaneswar, Odisha, India
palok.aich@niser.ac.in
[2] Homi Bhabha National Institute, Mumbai, India

Abstract. The intersection of metabolomics and microbiome research represents a critical frontier in understanding human health and disease. This work presents a systems-level approach integrating host-microbiome metabolic interactions, demonstrating how microbial communities influence health outcomes through metabolite-mediated communication. Using multi-omics integration, machine learning, and clinical validation studies, we establish a predictive framework for microbiome profiling from circulatory metabolites. Our research encompasses circadian metabolomics, aging and longevity studies, and translational applications in metabolic diseases including non-alcoholic steatohepatitis (NASH), Parkinson's disease (PD), obesity, and type 2 diabetes. The development of a novel gut function test and personalized probiotic interventions demonstrates the clinical utility of systems-level microbiome-metabolome analysis. These findings underscore the microbiome as a dynamic, responsive layer of human biology with significant implications for precision medicine and therapeutic innovation.

Keywords: metabolomics · microbiome · metabolic interactions · personalized probiotic interventions

1 Introduction

The human microbiome represents one of the most complex biological systems, forming a dynamic interface where microbial chemistry intersects with host metabolism [1]. The FDA Commissioner recently declared the microbiome "the most important frontier in medicine that we are not paying enough attention to" [2]. The microbiome-metabolome axis creates bidirectional communication networks influencing metabolic, immunological, and neurological health outcomes [3].

Historical analysis reveals that pre-colonial populations consumed one to two meals daily with integrated fasting practices, centered on locally sourced foods including millets, tubers, and lentils. Metabolic diseases were rare under these conditions. The colonial-era imposition of three-meal scheduling disrupted circadian rhythms and metabolic programming—a "metabolic colonization" contributing to contemporary disease epidemics. Modern research identifies seven convergent factors driving accelerated brain aging through gut-brain axis disruption: chronic inflammation, metabolic

G. B. Mohanty and V. S. Prabhudesai (Eds.): DAEC 2026, SPPHY 368, pp. 119–123, 2026.
https://doi.org/10.1007/978-981-92-2948-2_14

stress, environmental exposures, chronic stress, genetic vulnerability, and impaired neuroplasticity [4–6].

Large-scale studies demonstrate profound consequences of early microbiome disruption. Infant antibiotic exposure before age two increases risks for asthma, allergic disease, obesity, ADHD, and celiac disease [2]. Twin studies reveal that despite identical genomes, gut microbiomes show substantial variation influenced by environmental factors, emphasizing the microbiome's role as an environmentally responsive health determinant [7, 8].

2 Research Framework and Methodological Approaches

2.1 Multi-omics Integration and Experimental Framework

AichLab at NISER employs systems biology integrating metabolomics, metagenomics, and computational modeling. Research streams include divergent NASH phenotypes, metabolomic shifts in Parkinson's disease, circadian metabolomics, aging and longevity research, machine learning-based cross-disease insight, explainable AI-driven biomarker discovery, and microbe prediction from circulatory metabolites.

We implemented a circadian disruption protocol using C57BL/6 and BALB/c mice (male and female) subjected to light-dark (LD) or constant darkness (DD) conditions, sampled at weeks 3, 5, and 8. Untargeted LC-MS metabolomic profiling revealed extensive remodeling: 773 serum and 607 cecal metabolites under LD versus 972 serum and 833 cecal metabolites under DD. Metabolites were classified as microbiome-derived or host-derived based on biosynthetic pathways and metagenomic correlations.

16S rRNA sequencing demonstrated pronounced shifts in Firmicutes and Bacteroidetes phyla under DD conditions, with sex-specific and strain-specific responses. Correlation network analysis identified metabolite-microbe associations shifting under DD, with week-specific patterns distinguishing beneficial from harmful metabolite-microbe interactions. Lotka-Volterra-inspired mathematical modeling and phase portrait analysis formalized dynamic equilibria governing metabolite production and microbial abundances, enabling prediction of critical transition points.

2.2 Translational Applications and Clinical Validation

2.2.1 Gut Function Test and Clinical Translation

We developed a novel "Gut Function Test" predicting gut microbiome composition from blood metabolite profiles—a paradigm shift from stool-based to minimally invasive blood-based assessment. The workflow integrates: (1) cohort recruitment and sample collection, (2) untargeted LC-MS metabolomics and metagenomic sequencing, and (3) AI/ML algorithms selecting metabolite-microbe pairs to train predictive models. This enables species-level microbiome prediction from finger-prick blood samples or swabs, facilitating longitudinal monitoring.

Table 1. Clinical validation studies for gut function test and personalized probiotics

Condition	Institution	Status
Type 2 Diabetes, Obesity	AIIMS, Bhubaneswar	Completed
NAFLD	Aster Hospital, Bangalore	Completed
IBS, Stress and anxiety	MAMC, Delhi	Ongoing
Vaginal microbiome	PGIMS, Rohtak	Ongoing
Oral microbiome	IDS, Bhubaneswar	Ongoing
Multi-site validation	NISER, Bhubaneswar	Ongoing

2.2.2 Clinical Studies and Therapeutic Outcomes

Multiple clinical validation studies have been completed or are ongoing in collaboration with leading medical institutions (Table 1).

In the completed obesity cohort study, personalized probiotic interventions based on microbiome-metabolome profiling achieved approximately 8% mean weight loss. In the Type 2 Diabetes cohort, interventions resulted in approximately 40% reduction in postprandial blood glucose levels. These clinically significant outcomes validate the therapeutic potential of systems-level microbiome analysis and targeted intervention strategies.

2.2.3 Regulatory Approval and Commercialization

The gut function test platform has received regulatory approval and has been commercialized since October 2024 through MicrobioTx Health. Patent protection has been secured in India, with patents published in the United States and European Union. The technology represents the first and only predictive AI model to profile host microbiome at single-species resolution using blood or swab samples, with ongoing expansion to oral, vaginal, skin microbiomes and disorders including irritable bowel syndrome (IBS), inflammatory bowel disease (IBD), NAFLD, Parkinson's disease, stress-related disorders, and cancer.

The platform has gained recognition at national conferences, including the National Conference of Gastroenterology, and is supported by key opinion leaders in gastroenterology and metabolic medicine. This bench-to-business (B2B) translation exemplifies successful academic innovation reaching clinical practice.

3 Discussion and Future Perspectives

3.1 Systems Chemistry and Precision Medicine

Systems-level integration reveals the microbiome as a metabolic organ dynamically responding to host states, environmental inputs, and circadian rhythms. Disruption cascades through metabolic, inflammatory, and neurological pathways, with interventions at the microbiome-metabolome interface offering multi-system benefits for metabolic, cognitive, and longevity outcomes.

Blood-based microbiome profiling enables precision medicine previously constrained by sampling limitations. Longitudinal monitoring facilitates dynamic intervention adjustment, with demonstrated clinical efficacy in obesity (8% weight loss) and Type 2 Diabetes (40% glucose reduction). Cohort-specific probiotics represent personalized therapeutics tailored to individual microbiome deficiencies and metabolic imbalances.

Explainable AI (XAI) integration ensures predictive models provide mechanistic insight into metabolite-microbe interactions driving predictions, essential for clinical adoption and regulatory approval [9]. Mathematical modeling complements machine learning, enabling accurate prediction and causal inference [10].

Challenges include inter-individual variability requiring large diverse cohorts, temporal dynamics requiring longitudinal designs, and expansion to oral, vaginal, and skin microbiomes. Ethical considerations surrounding data privacy, equitable access, and regulatory frameworks require attention as microbiome medicine advances toward mainstream adoption.

4 Conclusion

The microbiome-metabolome interface represents a critical determinant of human health spanning metabolic, immunological, neurological, and longevity outcomes. Systems-level approaches integrating multi-omics profiling, computational modeling, and clinical validation enable both mechanistic insight and therapeutic translation. The development of blood-based microbiome profiling and personalized probiotic interventions demonstrates the feasibility and clinical utility of precision microbiome medicine. Circadian regulation emerges as a master coordinator of microbiome function, with disruption contributing to accelerated aging and multi-system disease. Addressing the "metabolic colonization" imposed by modern dietary and temporal patterns may offer population-level benefits for metabolic health. Continued innovation at the intersection of metabolomics, microbiomics, artificial intelligence, and clinical translation promises to establish the microbiome as a cornerstone of 21st-century medicine.

References

1. Aminian-Dehkordi, J., Montazeri, F., Tamadon, A., Mofrad, M.R.K.: Systems biology and microbiome innovations for personalized diabetic retinopathy management. npj Syst. Biol. Appl. **11**, 133 (2025)
2. Aversa, Z., et al.: Association of infant antibiotic exposure with childhood health outcomes. Mayo Clin. Proc. **96**, 66 (2020)
3. Sabih Ur Rehman, S., et al.: Integrative systems biology approaches for analyzing microbiome dysbiosis and species interactions. Brief. Bioinform. **26**(4), bbaf323 (2025)
4. Chatterjee, G., et al.: Microbiome systems biology advancements for natural well-being. Sci. Total Environ. **838**, 155915 (2022)
5. Teichman, E.M., O'Riordan, K.J., Gahan, C.G.M., Dinan, T.G., Cryan, J.F.: When rhythms meet the blues: circadian interactions with the microbiota-gut-brain Axis. Cell Metab. **31**, 448–471 (2020)
6. Su, J., Zhao, L., Fu, R., Tang, Z.: Linking circadian rhythms to gut-brain Axis lipid metabolism associated with endoplasmic reticulum stress in Alzheimer's disease. CNS Neurosci. Ther. **31**, e70329 (2025)

7. Xie, H., et al.: Shotgun metagenomics of 250 adult twins reveals genetic and environmental impacts on the gut microbiome. Cell Syst. **3**, 572–584.e3 (2016)
8. Bowyer, R.C.E., et al.: Socioeconomic status and the gut microbiome: a twins UK cohort study. Microorganisms. **7**, 17 (2019)
9. Wang, T., et al.: Predicting metabolomic profiles from microbial composition through neural ordinary differential equations. Nat. Mach. Intell. **5**, 284–293 (2023)
10. Kirtipal, N., Seo, Y., Son, J., Lee, S.: Systems biology of human microbiome for the prediction of personal glycaemic response. Diab. Metab. J. **48**, 821–836 (2024)

Epigenetic Regulators as Drivers of Tumor Heterogeneity

Chandrima Das[1,2(✉)]

[1] Biophysical Sciences Division, Saha Institute of Nuclear Physics, 1/AF Bidhannagar, Kolkata 700064, India
chandrima.das@saha.ac.in
[2] Homi Bhabha National Institute, Training School Complex, Anushakti Nagar, Mumbai 400094, India

Abstract. The epigenome functions as a critical regulator of cellular physiology through dynamic, reversible modifications that govern gene expression. Our laboratory investigates chromatin readers/effectors, which recognize specific epigenetic landscapes and regulate gene expression programs. Our findings demonstrate these proteins' pivotal roles in regulating cancer hallmarks, offering insights into tumor heterogeneity. The tumor microenvironment significantly influences cancer progression through interactions between resident cells and the extracellular matrix (ECM). It is shown that epigenetic regulators program ECM gene expression, thereby altering ECM stiffness and leading to the release of cell-free DNA via exosomes from cancer cells. Further, immune signaling pathways are also deregulated by the epigenetic factors directly impacting the metastatic potential of the cancer cells. Our research illuminates how specific chromatin readers interpret the epigenetic code in cancer, emerging as promising therapeutic targets for future interventions.

Keywords: Tumor microenvironment · tumor heterogeneity · metastatic potential · extracellular matrix remodeling · epigenetic regulation

1 Introduction

The underlying epigenetic regulations that play a pivotal role in fine-tuning gene expression programs in human diseases, with a special emphasis on cancer, are being investigated. Central to our research is the study of chromatin reader/effectors, particularly transcription factors with PHD finger and bromodomain modules, a class of proteins that recognize specific epigenetic landscapes and differentially regulate gene expression programs under physiological and pathophysiological conditions.

Cancer is a complex, multi-factorial disease. Along with the cancer cells, the tumor microenvironment is composed of multiple other cell types, viz., immune cells, stromal cells, as well as different structural components contributing to the extracellular matrix stiffness [1, 2]. There is distinct cellular heterogeneity reflected in early vs. late tumors. E.g., the Cancer Associated Fibroblasts remain quiescent in the early stage and become

G. B. Mohanty and V. S. Prabhudesai (Eds.): DAEC 2026, SPPHY 368, pp. 124–129, 2026.
https://doi.org/10.1007/978-981-92-2948-2_15

activated in the late stage (establishing apCAF, iCAF, myCAF lineages) [3]. Along with the Fibroblast cell heterogeneity, endothelial cell heterogeneity is observed in late tumors. Thus, the tumor microenvironment has a potential role in deciding the potential secondary metastatic niches.

2 Research Highlights

The following aspects were highlighted in the conclave.

Classical mechanism of epigenetic reprogramming in cancer: Epigenetic regulators serve as master orchestrators of gene expression, capable of reprogramming cancer cell identity across multiple oncogenic dimensions. A chromatin-reader protein, recognizing specific combinatorial histone modification signatures, exemplifies this multifaceted regulatory capacity [4]. By suppressing non-coding isoforms while reinforcing protein-coding transcripts essential for differentiation, these epigenetic regulators directly modulate cellular fate commitment [5]. Remarkably, these factors operate as a tumor suppressor in breast malignancies, reinstating differentiation programs that render cancer cells therapy-sensitive, thereby positioning epigenetically driven differentiation therapy as a non-toxic therapeutic modality [6].

Beyond differentiation, the epigenetic factors actively restrain tumor aggressiveness by suppressing epithelial-to-mesenchymal transition, a pivotal mechanism utilised by metastatic cancer cells. Through selective enrichment on promoters of epithelial maintenance genes, these factors preserve the cellular architecture, thereby limiting metastatic dissemination. Their antiproliferative functions further complement this anti-invasive role, collectively curtailing tumor growth in preclinical models. Additionally, these epigenetic regulators interface with chromatin remodelling complexes and transcription initiation machinery to broadly reshape the gene expression landscapes in cancer [7]. Interestingly, the telomere length maintenance is another hallmark of cancer cells. The post-translational modification of a telomerase catalytic subunit by a polycomb-associated enzyme triggers invasiveness by epigenetically silencing adhesion gene promoters, coupling telomere maintenance with EMT induction, revealing how convergent epigenetic mechanisms coordinately drive cancer progression and therapeutic resistance [8].

Epigenetic regulation of the metabolic landscape of tumor cells and the tumor-immune cross-talk: Cancer cells can thrive in nutrient-limiting and hypoxic environments and hence need to turn on metabolic programs for their sustenance. Under hypoxic conditions, tumor cells shift their energy metabolism toward anaerobic glucose breakdown, generating lactate that remodels the surrounding microenvironment. This metabolic adaptation is not merely a passive response but is actively orchestrated through epigenetic mechanisms, wherein chromatin-associated regulators modulate the transcription of key metabolic enzymes. Consequently, the immune surveillance gets suppressed, enabling tumors to evade cytotoxic immune responses. Targeting this epigenetically programmed metabolic axis, therefore, presents a compelling therapeutic strategy against aggressive malignancies [9] (Fig. 1).

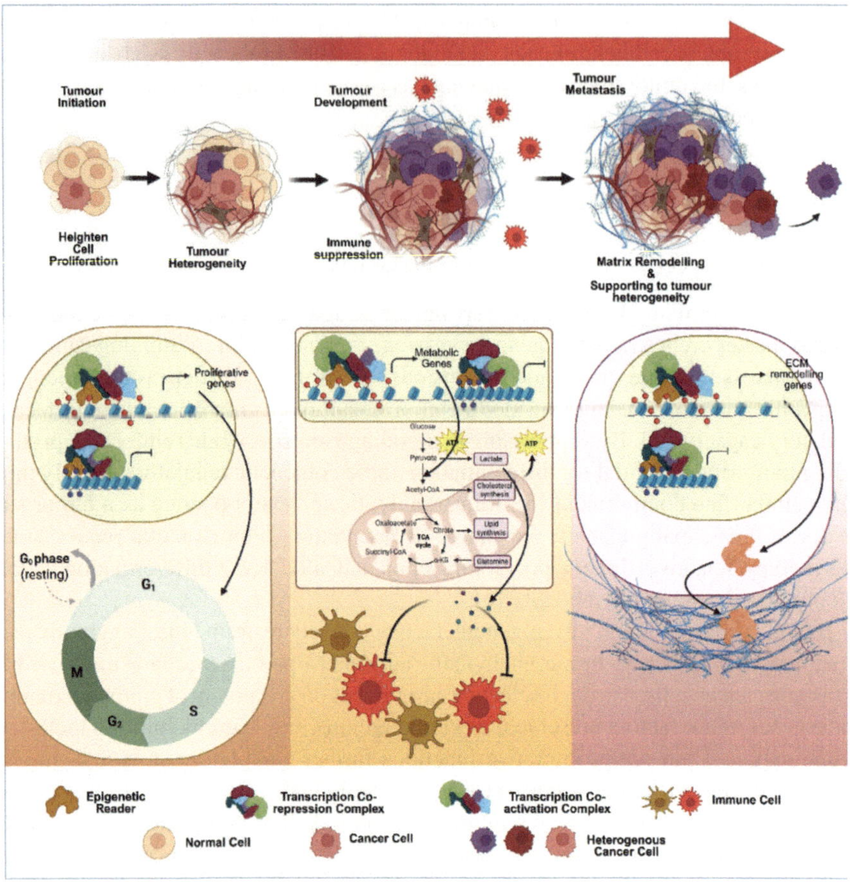

Fig. 1. Schematic illustration highlighting the key contributions to cancer research. Cancer is a multi-stage, progressive disease that originates from uncontrolled cell division and advances to metastasis, where malignant cells migrate and establish secondary tumors at distant sites. We investigate the role of epigenetic reader proteins in regulating key oncogenic processes. Our work focuses on elucidating the molecular mechanisms underlying cancer cell proliferation during tumor initiation, immune evasion during disease progression, and extracellular matrix remodeling and tumor micro-environment contributions during metastatic dissemination. Created in BioRender. Das, C. (2026) https://BioRender.com/2stj0ci

Contribution of the Extracellular Matrix in tumor metastasis from an epigenetic perspective:

The extracellular matrix serves as a dynamic structural scaffold that profoundly shapes tumor behavior beyond mere mechanical support. In aggressive breast malignancies, remodeling of this matrix architecture, characterized by altered collagen crosslinking activity of lysyl oxidases and enhanced deposition of fibrous protein, including collagen and fibronectin, contributes to the enhanced matrix stiffness. All these factors actively drive invasiveness and distant metastasis. Epigenetic factors play a seminal role in the modulation of the Epithelial-Mesenchymal-Transition by altering the transcription programs of the epithelial genes [10]. Beyond this classical mechanism, they are identified as crucial regulators of transcriptional programs encoding matrix-stiffening components. The resulting fibrotic microenvironment amplifies pro-invasive signaling cascades, facilitating tumor dissemination. The ECM stiffness modulates the cell-free DNA release from the exosomes of the cancer cells and can have a broader implication in setting up a pre-metastatic niche. Epigenetically restoring matrix homeostasis, therefore, emerges as a rational avenue for curtailing metastatic progression in refractory breast cancers [11] (Fig. 1).

Sensitization of the therapy-resistant cancer cells by the epigenetic factors: Besides working on the core functional networks in cancer cells that are impacted by epigenetic regulators, the pressing problem of therapeutic resistance is being investigated. Notably, the oncogenes maintained in poised epigenetic states may be activated during chemotherapy resistance. The ability of the epigenetic readers to reinstate the repression state through their potential interacting partners, sensing bivalent chromatin landscapes, may lead to chemosensitization [12]. Further, the ECM remodeling mechanism guided by alteration in the epigenetic landscape of chromatin also plays an important role in therapy resistance. Notably, certain ATRA-responsive chromatin readers have been identified, offering prospects in retinoid-based therapies [13, 14].

Epigenetic-Metabolic regulatory axis in cancer therapy resistance: Epigenetic reprogramming of cellular metabolism represents a central and emerging hallmark of cancer, profoundly influencing tumor progression, therapeutic resistance, and relapse. Cancer cells exploit epigenetically remodeled metabolic pathways to sustain malignant phenotypes, evade chemotherapy, and dynamically regulate DNA damage repair mechanisms, collectively reinforcing treatment failure. Metabolic heterogeneity, orchestrated through altered signaling networks and chromatin-level regulation, further complicates therapeutic intervention. Integrating genome-wide epigenetic analysis with metabolic profiling offers a powerful strategy to decode subtype-specific vulnerabilities. Ultimately, mechanistically targeting the epigenetics-metabolism-DNA repair axis holds considerable promise for sensitizing resistant tumors to conventional therapies and enabling precision oncology. These crucial aspects have been covered in the published reviews from our group [15–17].

3 Other Research Activities of the Group:

Apart from the highlighted discussions that have been described above, our group also focuses extensively on two other aspects- (i) Epigenetic reprogramming during Hepatitis B virus infection promoting hepatocellular carcinoma: During HBV infection, epigenetic regulators are hijacked to subvert host immunity and reprogram cellular metabolism, collectively driving hepatocellular carcinoma. The epigenetic regulator, resident in the PML Nuclear bodies, undergoes post-translational modification changes, enabling viral proteins to co-opt transcriptional machinery, suppress interferon-mediated antiviral defenses, and deregulate gluconeogenic gene expression through histone modification-dependent promoter remodeling, ultimately facilitating viral persistence and tumor progression [18–20] (ii) Epigenetic Reprogramming of Metabolic

Homeostasis: From Nutritional Stress to Metabolic Dysfunction and Hepatocellular Carcinoma: An epigenetic reader protein, recognizing active chromatin marks, serves as a versatile metabolic regulator across nutritional stress conditions. It suppresses gluconeogenesis by recruiting chromatin-remodeling co-repressors, coordinates glycolytic and mitochondrial energy balance alongside a tumor suppressor partner, and modulates lipid elongation pathways under fatty acid overload. Its dysregulation accelerates steatosis, inflammation, and fibrosis progression, collectively linking epigenetically programmed transcriptional responses to type 2 diabetes onset, fatty liver disease advancement, and hepatocellular carcinoma development [21–25] (Fig. 1).

4 Future Perspective

At present, efforts are being made to move beyond bulk RNA-seq by integrating single-cell and spatial transcriptomics approaches to dissect tumor heterogeneity at enhanced resolution. In parallel, epitherapies and epidrugs are being explored to comprehend their impact on gene-regulation dynamics. The establishment of these advanced capabilities at SINP is currently in progress. Additionally, patient-derived organoid models are being generated to ask more clinically relevant questions.

Acknowledgments. The research has been supported by "Basic and Applied Research in Biophysical Sciences" (RSI 4008) and "Basic and Applied Research in Biophysics and Material Science" (RSI 4002) by the Department of Atomic Energy, Govt. of India. We are also thankful for the research support from BT/PR52628/MED/30/2553/2024 by the Department of Biotechnology, SwarnaJayanti Fellowship (DST/SJF/LSA-02/2017–18), and Core Research Grant (CRG/2022/005052) by the Department of Science and Technology, Govt. of India, provided to CD.

References

1. Mai, Z., et al.: Modulating extracellular matrix stiffness: a strategic approach to boost cancer immunotherapy. Cell Death Dis. **15**, 307 (2024)
2. de Visser, K.E., et al.: The evolving tumor microenvironment: from cancer initiation to metastatic outgrowth. Cancer Cell **41**(3), 374–403 (2023)
3. Yang, D., et al.: Cancer-associated fibroblasts: from basic science to anticancer therapy. Exp. Mol. Med. **55**(7), 1322–1332 (2023)
4. Adhikary, S., et al.: Selective recognition of H3.1K36 Dimethylation/H4K16 acetylation facilitates the regulation of all-trans-retinoic Acid (ATRA)-responsive Genes by Putative Chromatin Reader ZMYND8. J. Biol. Chem. **291**(6):2664–2681 (2016)
5. Adhikary, S., et al.: ZMYND8 suppresses MAPT213 LncRNA transcription to promote neuronal differentiation. Cell Death Dis. **13**, 766 (2022)
6. Mukherjee, S., et al.: A novel role of tumor suppressor ZMYND8 in inducing differentiation of breast cancer cells through its dual-histone binding function. J. Biosci. **45**, 2 (2020)
7. Basu, M., et al.: Dual histone reader ZMYND8 inhibits cancer cell invasion by positively regulating epithelial genes. Biochem. J. **474**(11), 1919–1934 (2017)
8. Sanyal, S., et al.: SUMO E3 ligase CBX4 regulates hTERT-mediated transcription of CDH1 and promotes breast cancer cell migration and invasion. Biochem. J. **477**(19), 3803–3818 (2020)

9. Nandi, S., et al.: The hypoxia-induced chromatin reader ZMYND8 drives HIF-dependent metabolic rewiring in breast cancer. J. Biol. Chem. **301**(10), 110680 (2025)
10. Adhikary, S., et al.: Atypical plant homeodomain of UBR7 functions as an H2BK120Ub ligase and breast tumor suppressor. Nat. Commun. **10**(1), 1398 (2019)
11. Adhikari, S., et al.: UBR7 in concert with EZH2 inhibits the TGF-β signaling leading to extracellular matrix remodeling. Cell Rep. **43**(7), 114394 (2024)
12. Mukherjee, S., et al.: Suppression of poised oncogenes by ZMYND8 promotes chemo-sensitization. Cell Death Dis. **11**(12), 1073 (2020)
13. Basu, M., et al.: Chromatin reader ZMYND8 is a key target of all trans retinoic acid-mediated inhibition of cancer cell proliferation. Biochim. Biophys Acta Gene Regul Mech. **1860**(4), 450–459 (2017)
14. Guha, D., et al.: ZMYND8 is a regulator of sonic hedgehog SIGNALING in ATRA-Mediated differentiation of neuroblastoma cells. Biochemistry **63**(12), 1534–1542 (2024)
15. Adhikari, S., et al.: The paradigm of drug resistance in cancer: an epigenetic perspective. Biosci Rep. **42**(4):BSR20211812 (2022)
16. Das, C., et al.: Epigenetic-metabolic interplay in the DNA damage response and therapeutic resistance of breast cancer. Cancer Res. **83**(5), 657–666 (2023)
17. Das, C., et al.: A prismatic view of the epigenetic-metabolic regulatory axis in breast cancer therapy resistance. Oncogene **43**(23), 1727–1741 (2024)
18. Sengupta, I. et al.: Host transcription factor Speckled 110 kDa (Sp110), a nuclear body protein, is hijacked by hepatitis B virus protein X for viral persistence. **292**(50):20379–20393 (2017)
19. Sengupta, I., et al.: Epigenetic regulation of Fructose-1,6-bisphosphatase 1 by host transcription factor Speckled 110 kDa during hepatitis B virus infection. FEBS J. **289**(21), 6694–6713 (2022)
20. Singh, V., et al.: UBR7 E3 ligase suppresses interferon-β Mediated immune signaling by targeting Sp110 in Hepatitis B Virus-induced hepatocellular carcinoma. ACS Infect Dis. **10**(11), 3775–3796 (2024)
21. Sen, S., et al.: Transcription factor 19 interacts with histone 3 lysine 4 trimethylation and controls gluconeogenesis via the nucleosome-remodeling-deacetylase complex. J. Biol. Chem. **292**(50), 20362–20378 (2017)
22. Mondal, P., et al.: TCF19 promotes cell proliferation through binding to the histone H3K4me3 Mark. Biochemistry **59**(4), 389–399 (2020)
23. Mondal, P., et al.: TCF19 and p53 regulate transcription of TIGAR and SCO2 in HCC for mitochondrial energy metabolism and stress adaptation. FASEB J. **35**(9), e21814 (2021)
24. Mondal, A., et al.: TCF7l2 regulates fatty acid chain Elongase HACD3 during Lipid-Induced Stress. Biochemistry **64**(8), 1828–1840 (2025)
25. Mondal, A. et al.: Transcription factor 19 modulates fatty acid elongation and unfolded protein response to attenuate palmitic acid-induced hepatic dysfunction. Nat Commun. **17**(1), 5596 (2026)

Science-Based Perspectives on India's Ecology: Past, Present and Future

Uma Ramakrishnan(✉)

National Centre for Biological Sciences, TIFR, Bangalore, India
uramakri@ncbs.res.in

Abstract. India, although limited in land area, supports exceptional biodiversity and ecosystem services vital for human well-being and climate stability, but faces severe declines due to habitat loss and fragmentation. Over the last two decades, DAE/TIFR research has revealed that while Indian tigers retain high genetic diversity, their persistence depends on maintaining landscape connectivity. DAE/TIFR research also reveals that many so-called degraded forests are actually ancient savanna ecosystems, necessitating revised conservation priorities. There is a strong need for linking ecological monitoring, climate research, and conservation—to enhance ecosystem resilience. As scientists of TIFR/DAE, we hope to initiate and contribute strategic research to enable a science-based future for India's biodiversity and ecosystems.

Keywords: Conservation · ecosystems · India · climate vulnerability · long term monitoring

1 Introduction

Today, India is the most populous country in the world. Surprisingly, despite accounting for 2.3% of global land area, the Indian landmass is home to a disproportionate 8% of global biodiversity. It also hosts four of the earth's 35 biodiversity hotspots, and is one of eight global centres for crop diversity. India's unique and diverse ecosystems, distributed across our landscapes, rivers and oceans, are critical to our quality of life—and even to the survival of humanity. Nature and biodiversity are critical to humanity because they provide many services (see Fig. 1).

India's biodiversity and the associated natural services are valued at Rs.128 trillion/year from forests alone, underpin India's agricultural productivity, can help mitigate climate change, and can provide the foundation for a new model of economic and human development.

Unfortunately, over the last three centuries, India has lost significant natural habitats, which have been replaced by farmlands and human settlements. Figure 2 reveals how landscapes in India have changed since 1700, and reveal this intense fragmentation of habitat and land conversion.

The loss and fragmentation of natural habitats, direct exploitation of species and populations, and several other factors such as climate change, invasive species, pollution

G. B. Mohanty and V. S. Prabhudesai (Eds.): DAEC 2026, SPPHY 368, pp. 130–135, 2026.
https://doi.org/10.1007/978-981-92-2948-2_16

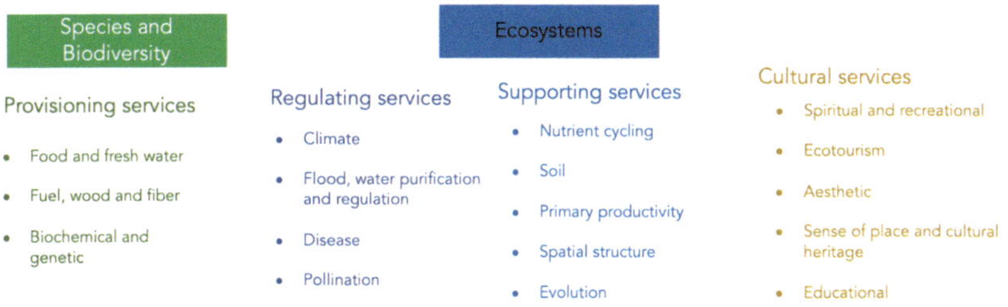

Fig. 1. Nature's contribution to people and their wellbeing can be detailed through varied biodiversity and ecosystem services

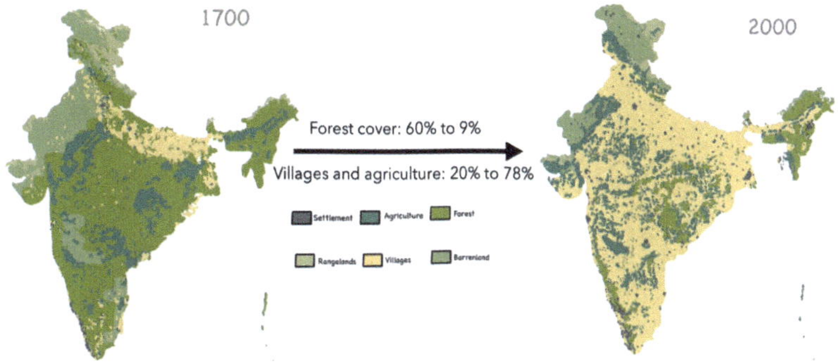

Fig. 2. Inferred land use landcover change between 1700 and 2000

and so on have two main consequences: the loss of biodiversity and the disruption of ecosystems, and the associated services. The loss of biodiversity also disrupts our future: biodiverse, functional ecosystems are the best way to mitigate the detrimental effects of climate change. Addressing this massive challenge requires (a) species conservation, (b) understanding ecosystems and associated services and (c) managing how species and ecosystems will respond to the effects of global climate change in order to curb biodiversity loss and mitigate the negative effects of climate change. In this article, I will explore examples of research from the DAE/TIFR systems that have contributed to conservation, ecosystem science, and propose a way forward.

2 Tiger Conservation Through a Genetic Lens

Like large carnivore species worldwide, tiger populations have declined historically, and now retain only 5% of its historical range. About 75% of the world's remaining 6,000 or so wild tigers inhabit the Indian subcontinent, making it an important region for attention to its conservation. Early on, we sampled tigers across the Indian subcontinent for mitochondrial and nuclear genetic variation, and revealed that Indian tigers retain more than half of the extant genetic diversity in the species. This diversity is retained despite a precipitous, likely human-induced population crash ~ 200 years ago in India [1]. Rapid development of genomic technologies allows us to now look at genetic variation at the scale of the genome, and a range-wide study of tiger genomes revealed similar results: Indian tigers have high total genomic variation, but also highlighted the substantial effects of recent habitat fragmentation on Indian tiger genomes [2]. These genomic studies also helped identify small, isolated tiger populations (in Rajasthan and Odisha, and the effects such isolation is having. For example, tigers in Ranthambore, Rajasthan are twice as inbred as tigers in other parts of India [3]. Further, isolation has led to high frequency of a rare phenotype, the black tiger, in Similipal Tiger Reserve [4]. Given the extreme isolation, inbreeding, and low variation in Similipal tigers, we suggested genetic supplementation for this population. These suggestions were implemented by managers, resulting in the translocation of two females from Tadoba Andhari Tiger Reserve in Maharashtra (Fig. 3).

Fig. 3. A black tiger in Similipal Tiger reserve (Photo from creative commons)

Clearly, gene flow or connectivity between tiger populations is important. We investigated factors that contribute to geneflow and connectivity in current populations, landscape genetics approaches to investigate correlations between genetic data (collected from Central India) and resistance to movement offered by landscape features (human settlements, forest cover, roads). We showed that human footprints on the landscape (such as high traffic roads, urban settlements) significantly impact recent movement of tigers across this landscape [5, 6]. Interestingly, roads and human footprint impact the varied species in this landscape (jungle cat, leopard, tiger, sloth bear) differentially, with smaller species being least impacted [7].

We then used simulations to explore how best to sustain tiger connectivity in the future within the Central Indian landscape [6]. Such simulations identified which populations are more likely to go extinct, but also indicated how to minimize extinction probability in the coming century. Our results suggest that securing corridors, with new protected sites within them is the way ahead. These results were used in a public interest litigation that resulted in Asia's longest underpass for wildlife.

In summary, while the attention of conservation biologists and governments has largely been directed at the national population size of the tigers, research at TIFR unequivocally establishes that connectivity between populations that make up these national numbers is critical for sustaining tigers and their diversity.

3 A Misunderstood Ecosystem: The Indian Savana

Biomes are large, naturally occurring sets of plants and animals that together occupy a major habitat. Early ecologists suggested that biomes were characterized by their distinctive climate, vegetation, and animal life. In India, most of the habitat was thought to be part of the dry deciduous biome, and grasslands were thought to be degraded, and hence were undervalued in conservation and habitat management. The savanna biome, defined as discontinuous tree cover in a continuous grass understory, was thought to be absent from India.

In pioneering research, TIFR scientists [8] investigated the continuum between forests and savannas. They asked a simple question: in intermediate rainfall regimes, is a habitat a degraded forest or is it actually a savanna? And how can one tell? They suggested that intermediate rainfall savannas should be dominated by fire tolerant and shade intolerant species. In contrast, forests should be characterized by fire intolerant and shade tolerant species. Given these differences, they were able to suggest diagnostic morphological, physiological and life-history traits that should characterise vegetation species in forests versus savannas. This led them to the observation that many habitats in South Asia are misclassified as forests, while they should be recognized as savannas.

These ideas led to further data-based focus and investigation of Asian savannas [9]. They highlighted paleontological records that reveal that Asian savannahs are over a million years old. They then used savanna climate domains from other continents to map putative savannas in South Asia. These analyses revealed that Asia has many climatic spaces very similar to African savannas. They then identified threats that Asia's savannas face, given that they are misunderstood to be forests.

Detailed characterization of savanna biomes from various continents revealed that the climate spaces true savannas occupy differs across continents [10]. These various research endeavors suggest that even in the most basic sense, we do not have a good idea of Indian habitats and ecosystems. Going forward, long term monitoring of vegetation and its dynamics is critical to better understand Indian ecosystems. In an attempt to build such knowledge, TIFR scientists launched LEMoN, the Long term Ecosystem Monitoring Network (https://lemonindia.weebly.com/).

4 The Way Ahead: Securing India's Ecological Future

Natural and managed ecosystems in India provide a variety of ecosystem services to people at local, regional, and national scales. Biodiversity underpins all of these ecosystem services, and their resilience to climate and other global change drivers. India's ecological and economic security is contingent on the continued ability of our ecosystems to provide irreplaceable services that are necessary for the well-being and sustainable economic development of its society. However, widespread biodiversity loss, invasions by exotic species, land degradation as a result of mismanagement, and land-use and climate change is threatening the integrity of our ecosystems and erode their ability to continue providing vital ecosystem services on which all of us are dependent (e.g. ref 11–14). Average global temperatures have increased 1.1 °C since the preindustrial times—faster since 1970 than in any preceding 50-year period over at least the last 2000 years—and set to rise further without drastic remedial actions [15]. In the face of a warming climate, heightened nutrient deposition, habitat loss and overexploitation, species and ecosystems across the globe face rapid changes that may reorganize ecological communities and systems. *In India, these processes remain unknown, although their impacts are predicted to be detrimental to both biodiversity and people, given that high human population density, intense land use and ongoing land cover change closely overlap with biodiversity-rich ecosystems.*

In the future, TIFR/DAE institutions hope to comprehend how climate is changing, how individuals, species, communities and ecosystems will respond to these changes, how to use monitoring data to predict trajectories, and inform restoration and conservation in varied ecosystems, including the marine realm. Finally, understanding and enabling climate literacy amongst the public and children, and high quality and well-trained researchers in climate change will be critical to maintain a science-based approach to manage global environmental burdens, enable behavioral change, and move towards a sustainable future. We hope for a cross-disciplinary, multi-centric, virtual platform to address core scientific issues around global environmental changes (including climate), with potential societal impact.

References

1. Mondol, S., Karanth, K.U., Ramakrishnan, U.: Why the Indian subcontinent holds the key to global tiger recovery. PLoS Genet. **5**(8), e1000585 (2009)
2. Armstrong, E.E., Khan, A., Taylor, R.W., Gouy, A., Greenbaum, G., Thiéry, A., Kang, J.T., Redondo, S.A., Prost, S., Barsh, G., Kaelin C.: Recent evolutionary history of tigers highlights contrasting roles of genetic drift and selection. Mol. Biol. Evol. **38**(6):2366–2379 (2021)
3. Khan, A., et al.: Genomic evidence for inbreeding depression and purging of deleterious genetic variation in Indian tigers. Proc. Natl. Acad. Sci. **118**(49), e2023018118 (2021)
4. Sagar, V., et al.: High frequency of an otherwise rare phenotype in a small and isolated tiger population. Proc. Natl. Acad. Sci. **118**(39), e2025273118 (2021)
5. Joshi, A., Vaidyanathan, S., Mondol, S., Edgaonkar, A., Ramakrishnan, U.: Connectivity of tiger (Panthera tigris) populations in the human-influenced forest mosaic of central India. PLoS ONE **8**(11), e77980 (2013)
6. Thatte, P., Joshi, A., Vaidyanathan, S., Landguth, E., Ramakrishnan, U.: Maintaining tiger connectivity and minimizing extinction into the next century: insights from landscape genetics and spatially-explicit simulations. Biol. Conserv. **218**:181–191 (2018)
7. Thatte, P., et al.: Human footprint differentially impacts genetic connectivity of four wide-ranging mammals in a fragmented landscape. Divers. Distrib. **26**(3), 299–314 (2020)
8. Ratnam, J., et al.: When is a 'forest'a savanna, and why does it matter? Glob. Ecol. Biogeogr. **20**(5), 653–660 (2011)
9. Ratnam, J., Tomlinson, K.W., Rasquinha, D.N., Sankaran, M.: Savannahs of Asia: antiquity, biogeography, and an uncertain future. Philos. Trans. R. Soc. B: Biol. Sci. **371**(1703) (2016)
10. Lehmann, C.E., et al.: Savanna vegetation-fire-climate relationships differ among continents. Science **343**(6170), 548–552 (2014)

11. Naeem, S., Thompson, L.J., Lawler, S.P., Lawton, J.H., Woodfin, R.M.: Declining biodiversity can alter the performance of ecosystems. Nature **368**(6473), 734–737 (1994)
12. Chapin Iii, F.S., Zavaleta, E.S., Eviner, V.T., Naylor, R.L., Vitousek, P.M., Reynolds, H.L., Hooper, D.U., Lavorel, S., Sala, O.E., Hobbie, S.E., Mack M.C.: Consequences of changing biodiversity. Nature. **405**(6783):234–242 (2000)
13. Dobson, A., Lodge, D., Alder, J., Cumming, G.S., Keymer, J., McGlade, J., Mooney, H., Rusak, J.A., Sala, O., Wolters, V., Wall, D.: Habitat loss, trophic collapse, and the decline of ecosystem services. Ecology. **87**(8):1915–1924 (2006)
14. Isbell, F., et al.: Biodiversity increases the resistance of ecosystem productivity to climate extremes. Nature **526**(7574), 574–577 (2015)
15. https://www.ipcc.ch/report/ar6/syr/downloads/report/IPCC_AR6_SYR_SPM.pdf

Clinical Cancer Research in Resource-Limited Settings: Lessons Learned and Innovations from India

Vanita Noronha(✉)

Department of Medical Oncology, Tata Memorial Hospital, Parel, Mumbai, India
vanita.noronha@gmail.com

Abstract. Cancer care in low- and middle-income countries (LMICs) is shaped as much by health system constraints as by tumor biology. Delayed diagnosis, fragmented care pathways, limited access to therapies, and financial toxicity frequently compromise delivery of optimal care and oncologic outcomes. In this context, clinical research from India has increasingly focused on developing strategies that are not only effective, but also feasible and scalable within existing resources. Across disease sites such as lung cancer and head and neck cancer, this work highlights the importance of optimizing treatment delivery alongside therapeutic innovation. In *EGFR*-mutant lung cancer, approaches such as combining tyrosine kinase inhibitors with chemotherapy offer a pragmatic alternative in settings where access to newer agents is limited. In head and neck cancer, neoadjuvant chemotherapy has enabled a subset of patients with borderline resectable disease to undergo curative surgery, while refinements in chemoradiation and exploration of alternative regimens have addressed gaps in cisplatin-ineligible populations. Metronomic chemotherapy, with or without low-dose immunotherapy, represents another important advance, demonstrating that meaningful clinical benefit can be achieved with lower toxicity and substantially reduced cost. Beyond drug-based strategies, system-level interventions such as patient navigation and structured symptom monitoring improve treatment adherence and reduce unplanned healthcare utilization. Collectively, these efforts underscore a broad redefinition of innovation in oncology; one that prioritizes accessibility, affordability, and real-world applicability. Lessons from such settings have relevance beyond LMICs, particularly as healthcare systems globally confront rising costs and inequities in access to care.

Keywords: Lung Cancer · Head and Neck Cancer · Clinical Cancer Research · Chemoradiation

1 Introduction

Looking beyond drugs: When we think about progress in oncology, the conversation often centers around new drugs, molecular targets, and breakthrough trials. Yet, for many patients across the world, particularly in low- and middle-income countries (LMICs), the

G. B. Mohanty and V. S. Prabhudesai (Eds.): DAEC 2026, SPPHY 368, pp. 136–142, 2026.
https://doi.org/10.1007/978-981-92-2948-2_17

barriers to care are far more basic. Delays in diagnosis, difficulty navigating the health-care system, interruptions in treatment, and the financial burden of therapy frequently shape outcomes as much as the disease itself [1].

In countries like India, these challenges are encountered daily. Patients often present with advanced disease, not necessarily because of aggressive biology alone, but because of systemic delays and access issues [2]. This raises an important question: are we focusing too narrowly on innovation in drugs, while underestimating the impact of improving how care is delivered?

Over the past several years, clinical research from India has increasingly tried to address this gap. The focus has been on developing strategies that are not only effective, but also realistic within existing constraints. This has led to a body of work that prioritizes access, affordability, and adaptability, alongside efficacy.

2 Lung Cancer: Making the Most of Available Therapies

The management of EGFR-mutated lung cancer has been transformed by tyrosine kinase inhibitors [3]. Early studies such as IPASS established that targeted therapy could significantly improve response rates and delay disease progression compared to chemotherapy [4]. Subsequent trials with other TKIs confirmed these benefits.

However, despite improvements in progression-free survival, gains in overall survival have been less striking. In routine practice, especially in resource-constrained settings, this gap becomes even more evident. Access to newer agents such as osimertinib is limited by cost, and many patients are unable to receive optimal sequential therapy.

This has prompted an alternative line of thinking: rather than relying solely on newer and more expensive drugs, can we improve outcomes by modifying how existing treatments are used?

One such approach has been the combination of first-generation TKIs with chemotherapy. While this strategy may appear less "cutting-edge," it is far more accessible and has the potential to improve outcomes in a larger proportion of patients. We found that in patients with EGFR-mutant non-small cell lung cancer, first line treatment with gefitinib and chemotherapy resulted in a statistically significant and meaningful improvement in both progression-free and overall survival [5], which persisted on long-term follow-up [6]. In settings where the cost of therapy directly determines whether a patient is treated at all, such approaches are not just pragmatic; they are essential.

This reinforces an important idea: a treatment strategy is only as effective as its accessibility in the real world.

3 Head and Neck Cancer: Expanding the Boundaries of Curative Treatment

Head and neck cancers, particularly those involving the oral cavity, remain a major health burden in India [7]. A significant proportion of patients present with locally advanced disease, where achieving a cure is challenging [8].

Traditionally, patients with technically unresectable tumors had limited options and poor outcomes [9]. The introduction of neoadjuvant chemotherapy has changed this landscape to some extent [10]. In a large cohort, a meaningful proportion of tumors could be downstaged sufficiently to allow surgery, and those who underwent resection had markedly better survival compared to those who did not [11–13].

What is striking here is not just the improvement in outcomes, but the shift in treatment intent: from purely palliative to potentially curative in a subset of patients who would otherwise have had very limited options.

At the same time, there has been growing interest in organ preservation strategies. While well established in some head and neck subsites, their role in oral cavity cancers has been less clear [14]. Emerging evidence suggests that, in carefully selected patients, it may be possible to preserve function without compromising survival, although this requires careful patient selection and remains an evolving area [15].

4 Refining Chemoradiation: The Value of Incremental Change

Concurrent chemoradiation continues to be the backbone of treatment for many locally advanced head and neck cancers [16]. Evidence from large meta-analyses has confirmed that adding chemotherapy to radiation improves survival, particularly when given concurrently [17]. However, even within this established framework, there is room for optimization. Questions around dosing schedules, for example, weekly versus three-weekly cisplatin have important practical implications. While three-weekly cisplatin may offer better disease control, it is also associated with greater toxicity, and not all patients are able to complete treatment as planned [18]. We tested whether lower dose cisplatin at 30 mg/m^2 given once a week along with radiation was non-inferior to standard high dose cisplatin at 100 mg/m^2 given once in three weeks. We found that low dose cisplatin at 30 mg/m^2 was not non-inferior to high dose cisplatin, which therefore should remain the standard of care [19, 20].

Similarly, efforts to enhance outcomes by adding targeted therapies have shown some promise, and the addition of nimotuzumab to cisplatin-based concurrent chemoradiation is particularly promising [21, 22].

An equally important issue is the management of patients who are not candidates for cisplatin. This is a group that is frequently encountered in clinical practice but underrepresented in trials. Studies exploring alternatives such as weekly docetaxel with radiation suggest that treatment can be tailored to these patients, although this often comes with trade-offs in terms of toxicity [23].

Taken together, these efforts highlight the importance of fine-tuning existing treatments, rather than relying solely on entirely new approaches.

5 Metronomic Therapy and Low-Dose Immunotherapy: Rethinking Intensity

One of the more distinctive contributions from LMIC-based research has been the development of metronomic chemotherapy [24]. This approach uses low, regular doses of oral agents, aiming to control disease while minimizing toxicity [25].

In advanced head and neck cancer, metronomic regimens have demonstrated outcomes comparable to conventional intravenous chemotherapy, with better tolerability and quality of life [26–28]. For patients who are frail, elderly, or unable to afford intensive treatment, this provides a meaningful alternative.

Building on this concept, recent studies have explored the addition of low-dose immunotherapy. The results are encouraging, with improvements in both progression-free and overall survival [29]. What makes this particularly relevant is the dramatic difference in cost compared to standard immunotherapy regimens.

This is not a trivial consideration. In many settings, only a small fraction of patients who could benefit from immunotherapy are actually able to receive it [30]. By reducing the dose and cost while retaining efficacy, these strategies offer a way to extend the benefits of modern oncology to a much larger population.

6 Strengthening Systems of Care

While drug-based innovations are important, they are only one part of the picture. Many of the challenges in cancer care arise from gaps in the system itself: missed appointments, delayed investigations, poor symptom control, and lack of coordination between different aspects of care [31].

Interventions such as patient navigation and structured symptom monitoring address these gaps directly. By helping patients move more efficiently through the system and by identifying problems early, these approaches can improve adherence to treatment and reduce unplanned hospital visits [32].

Although these interventions may not immediately translate into improved survival, they significantly enhance the quality and reliability of care delivery. In many ways, they represent low-cost, high-impact solutions that are highly relevant across diverse healthcare settings.

7 Conclusion: Broadening the Definition of Innovation

The experience from India underscores a broader lesson for oncology: innovation is not limited to new molecules or technologies. It also lies in rethinking how care is delivered, particularly in environments where resources are constrained. Three themes stand out.

First, treatment strategies must be grounded in the realities of the healthcare system in which they are applied. Evidence that does not account for context has limited value in practice.

Second, access and affordability are central to effectiveness. A therapy that cannot reach patients cannot improve outcomes, regardless of its efficacy in trials.

Finally, meaningful progress often comes from incremental changes: adjusting doses, combining existing treatments, or improving care delivery systems.

As healthcare systems worldwide grapple with rising costs and increasing complexity, these lessons are becoming increasingly relevant. The challenge ahead is not only to develop better treatments, but to ensure that they are accessible, sustainable, and applicable to the populations that need them most.

References

1. Nagar, A., Madamanchi, D., Nair, G.R., Revikumar, A., Ray, S., Vajjala, S.M., BS, A., Shivale, S.: Barriers to cancer diagnosis and treatment: a pilot qualitative study of patient and practitioner perspectives in rural India. Cureus **16**(8), e67249 (2024). https://doi.org/10.7759/cureus.67249
2. Datta, S.S., Ghose, S., Ghosh, M., Jain, A., Mandal, S., Chakraborty, S., Caduff, C.: Journeys: understanding access, affordability and disruptions to cancer care in India. Ecancermedicalscience **16**, 1342 (2022). https://doi.org/10.3332/ecancer.2022.1342
3. Sharma, M., Basu, D.A., Nathany, S., Amrith, B.P., Batra, U.: A narrative review of the role of common EGFR mutations in pathogenesis and treatment of non-small-cell lung carcinoma. Cancer Res., Stat., Treat. **5**(3), p507–518 (2022). https://doi.org/10.4103/crst.crst_222_22
4. Mok, T.S., et al.: Gefitinib or carboplatin-paclitaxel in pulmonary adenocarcinoma. N. Engl. J. Med. **361**(10), 947–957 (2009). https://doi.org/10.1056/NEJMoa0810699
5. Noronha, V., et al.: Gefitinib Versus Gefitinib Plus Pemetrexed and carboplatin chemotherapy in *EGFR*-mutated lung cancer. J. Clin. Oncol. **38**(2), 124–136 (2020). https://doi.org/10.1200/JCO.19.01154
6. Noronha, V., et al.: Gefitinib vs Gefitinib Plus pemetrexed and carboplatin chemotherapy in EGFR-variant lung cancer-long-term results of a randomized clinical trial. JAMA Oncol. **10**(6), 824–826 (2024). https://doi.org/10.1001/jamaoncol.2024.0584
7. Bagal, S., et al.: Head and neck cancer burden in India: an analysis from published data of 37 population-based cancer registries. Ecancermedicalscience **17**, 1603 (2023). https://doi.org/10.3332/ecancer.2023.1603
8. Fagan, J.J., Noronha, V., Graboyes, E.M.: Making the best of limited resources: improving outcomes in head and neck cancer. Am. Soc. Clin. Oncol. Educ. Book **41**, 1–11 (2021). https://doi.org/10.1200/EDBK_320923
9. Patil, V.M., Noronha, V., Muddu, V.K., Gulia, S., Bhosale, B., Arya, S., Juvekar, S., Chatturvedi, P., Chaukar, D.A., Pai, P., D'cruz, A., Prabhash, K.: Induction chemotherapy in technically unresectable locally advanced oral cavity cancers: Does it make a difference? Indian J. Cancer **50**(1), 1–8 (2013). https://doi.org/10.4103/0019-509X.112263
10. Goel, A., Singla, A., Prabhash, K.: Neoadjuvant chemotherapy in oral cancer: current status and future possibilities. Cancer Res. Stat. Treat. **3**(1), 51–59 (2020). https://doi.org/10.4103/CRST.CRST_79_19
11. Patil, V.M., et al.: Neoadjuvant chemotherapy followed by surgery in very locally advanced technically unresectable oral cavity cancers. Oral Oncol. **50**(10), 1000–1004 (2014). https://doi.org/10.1016/j.oraloncology.2014.07.015
12. Noronha, V., et al.: Phase 3 RCT comparing docetaxel-platinum with docetaxel-platinum-5FU as neoadjuvant chemotherapy in borderline resectable oral cancer. Eur. J. Cancer **200**, 113560 (2024). https://doi.org/10.1016/j.ejca.2024.113560
13. Noronha, V., et al.: Long-term outcomes of neo-adjuvant chemotherapy on borderline resectable oral cavity cancers: real-world data of 3266 patients and implications for clinical practice. Oral Oncol. **148**, 106633 (2024). https://doi.org/10.1016/j.oraloncology.2023.106633
14. Tschiesner, U.: Preservation of organ function in head and neck cancer. GMS Curr. Top Otorhinolaryngol Head Neck Surg. **11**, Doc07 (2012). https://doi.org/10.3205/cto000089
15. Chaukar, D., et al.: Prospective phase II open-label randomized controlled trial to compare mandibular preservation in upfront surgery with neoadjuvant chemotherapy followed by surgery in operable oral cavity cancer. J. Clin. Oncol. **40**(3), 272–281 (2022). https://doi.org/10.1200/JCO.21.00179

16. Brizel, D.M., Esclamado, R.: Concurrent chemoradiotherapy for locally advanced, nonmetastatic, squamous carcinoma of the head and neck: consensus, controversy, and conundrum. J. Clin. Oncol. **24**(17), 2612–2617 (2006). https://doi.org/10.1200/JCO.2005.05.2829
17. Lacas, B., et al.: MACH-NC collaborative group. meta-analysis of chemotherapy in head and neck cancer (MACH-NC): an update on 107 randomized trials and 19,805 patients, on behalf of MACH-NC group. Radiother Oncol. **156**, 281–293 (2021). https://doi.org/10.1016/j.radonc.2021.01.013
18. Mohamed, A., et al.: Concurrent chemoradiotherapy with weekly versus triweekly cisplatin in locally advanced squamous cell carcinoma of the head and neck: comparative analysis. Head Neck **41**(5), 1490–1498 (2019). https://doi.org/10.1002/hed.25379
19. Noronha, V., et al.: Once-a-week versus once-every-3-weeks cisplatin chemoradiation for locally advanced head and neck cancer: a phase III randomized noninferiority trial. J. Clin. Oncol. **36**(11), 1064–1072 (2018). https://doi.org/10.1200/JCO.2017.74.9457
20. Noronha, V., Patil, V., Menon, N., Singh, A., Shah, M., Shetake, A., Peelay, Z., Mathrudev, V., Nawale, K., Shah, S., Prabhash, K.: Long-term update of a phase 3 randomized study comparing once-a-week versus once-every-3-weeks cisplatin along with radiation in head and neck cancer. Int. J. Radiat. Oncol. Biol. Phys. **121**(1), 128–136 (2025). https://doi.org/10.1016/j.ijrobp.2024.07.2315
21. Patil, V.M., et al.: A randomized phase 3 trial comparing nimotuzumab plus cisplatin chemoradiotherapy versus cisplatin chemoradiotherapy alone in locally advanced head and neck cancer. Cancer **125**(18), 3184–3197 (2019). https://doi.org/10.1002/cncr.32179
22. Noronha, V., et al.: Nimotuzumab-cisplatin-radiation versus cisplatin-radiation in HPV negative oropharyngeal cancer. Oncotarget **11**(4), 399–408 (2020). https://doi.org/10.18632/oncotarget.27443
23. Patil, V.M., et al.: Results of phase III randomized trial for use of docetaxel as a radiosensitizer in patients with head and neck cancer, unsuitable for cisplatin-based chemoradiation. J. Clin. Oncol. **41**(13), 2350–2361 (2023). https://doi.org/10.1200/JCO.22.00980
24. Patil, V., Noronha, V., D'cruz, A.K., Banavali, S.D., Prabhash, K.: Metronomic chemotherapy in advanced oral cancers. J. Cancer Res. Ther. **8**(Suppl 2), S106–S110 (2012). https://doi.org/10.4103/0973-1482.92223
25. Sethuraman, S., Ramalingam, K.: Metronomic chemotherapy in oral cancer: a review. Cureus **15**(12):e49825 (2023). https://doi.org/10.7759/cureus.49825
26. Patil, V., et al.: Low-cost oral metronomic chemotherapy versus intravenous cisplatin in patients with recurrent, metastatic, inoperable head and neck carcinoma: an open-label, parallel-group, non-inferiority, randomised, phase 3 trial. Lancet Glob. Health **8**(9), e1213–e1222 (2020). https://doi.org/10.1016/S2214-109X(20)30275-8
27. Kothari, R., et al.: Phase 3 randomized study of physician choice vs metronomic chemotherapy in platinum refractory/ineligible head and neck cancer in palliative setting with survival outcomes. Cancer Lett. **621**, 217702 (2025). https://doi.org/10.1016/j.canlet.2025.217702
28. Kapoor, A., et al.: Triple oral metronomic chemotherapy versus chemotherapy of physician discretion after failure of platinum-based therapy in advanced head and neck cancer: a phase III randomized study (METRO-CHASE study). JCO Glob Oncol. **11**, e2500032 (2025). https://doi.org/10.1200/GO-25-00032
29. Patil, V.M., et al.: Low-dose immunotherapy in head and neck cancer: a randomized study. J. Clin. Oncol. **41**(2), 222–232 (2023). https://doi.org/10.1200/JCO.22.01015
30. Patil, V., et al.: Retrospective analysis: checkpoint inhibitor accessibility for thoracic and head and neck cancers and factors influencing it in a tertiary centre in India. Ecancermedicalscience. **16**, 1464 (2022). https://doi.org/10.3332/ecancer.2022.1464

31. Brand, N.R., Qu, L.G., Chao, A., Ilbawi, A.M.: Delays and barriers to cancer care in low- and middle-income countries: a systematic review. Oncologist **24**(12), e1371–e1380 (2019). https://doi.org/10.1634/theoncologist.2019-0057
32. Prabhash, K., et al.: A randomized trial of patient navigation with symptom-monitoring in advanced lung cancer. Eur. J. Cancer **228**, 115730 (2025). https://doi.org/10.1016/j.ejca.2025.115730

Radioisotopes for Healthcare

Rubel Chakravarty[1,2](✉)

[1] Radiopharmaceuticals Division, Bhabha Atomic Research Centre (BARC), Mumbai, India
rubelc@barc.gov.in
[2] Homi Bhabha National Institute, Anushaktinagar, Mumbai, India

Abstract. Radioisotopes play a pivotal role in modern healthcare by enabling highly sensitive molecular imaging and targeted radionuclide therapy. Over the past six decades, India has established a comprehensive programme under the Department of Atomic Energy (DAE) that integrates reactor- and accelerator-based radioisotope production with radiochemical processing and radiopharmaceutical development. Today, several clinically important radioisotopes and state-of-the-art radiopharmaceuticals are produced domestically. This chapter reviews the programme's evolution, highlights key indigenous innovations, and outlines its growing clinical impact in advancing theranostics and personalized nuclear medicine.

Keywords: Nuclear medicine · Radioisotopes · Radiopharmaceuticals · Targeted radionuclide therapy · Theranostics

1 Introduction

Radioisotopes have transformed modern medicine by making it possible to visualize biological processes and treat diseases at the molecular level [1]. In nuclear medicine, radiopharmaceuticals—biologically active molecules labelled with radioisotopes—accumulate selectively in organs or tumors, allowing clinicians to both diagnose and treat disease. Diagnostic radioisotopes enable imaging techniques such as positron emission tomography (PET) and single photon emission computed tomography (SPECT), while therapeutic radioisotopes deliver cytotoxic radiation directly to diseased tissues. The convergence of these two capabilities has led to the concept of theranostics, where closely related radioisotopes or certain radioisotopes possessing inherent physical properties enabling both diagnostic imaging (PET or SPECT) and targeted therapy (alpha or beta emission), are used for both imaging and therapy, enabling increasingly personalized approaches to cancer treatment [1]. Figure 1 provides an overview of the use of radioisotopes in nuclear medicine. In India, the development of radioisotope science and technology has been closely associated with the programmes of DAE. Over the decades, research reactors, accelerator facilities, radiochemical and radiopharmaceutical laboratories, biological evaluation facilities and clinical collaborations have together created a strong and largely self-reliant ecosystem for the production and application of medical radioisotopes.

G. B. Mohanty and V. S. Prabhudesai (Eds.): DAEC 2026, SPPHY 368, pp. 143–149, 2026.
https://doi.org/10.1007/978-981-92-2948-2_18

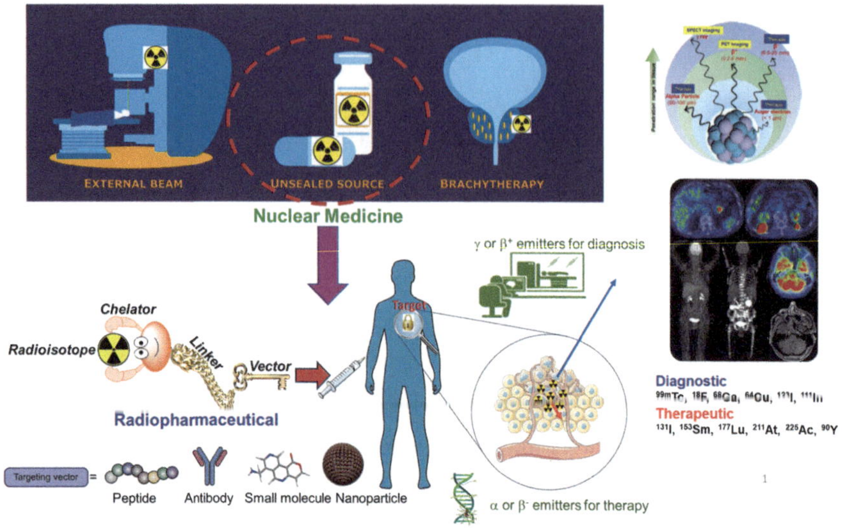

Fig. 1. Schematic illustration of the role of radioisotopes in modern medical practices (sealed sources for external beam therapy and brachytherapy and unsealed sources for use in nuclear medicine). Usually unsealed sources are administered intravenously in the form of radiopharmaceuticals, which are formed by labeling suitable biologically active molecules with radioisotopes, to enable molecularly targeted and non-invasive imaging using PET and SPECT, as well as targeted radionuclide therapy.

2 Evolution of the Radioisotope Program in India

India's radioisotope programme began with the establishment of the Atomic Energy Commission (AEC) in 1948 under the leadership of Dr. Homi Jehangir Bhabha, who championed the peaceful uses of atomic energy for national development. A major milestone followed in 1956 with the commissioning of APSARA reactor at the Atomic Energy Establishment, Trombay (AEET), later renamed as Bhabha Atomic Research Centre (BARC). APSARA enabled the irradiation of target materials for producing radioisotopes for scientific and medical applications. Early radiochemical processing was carried out at off-site laboratories in Mumbai, where radioisotopes such as ^{24}Na, ^{32}P, ^{42}K, and ^{131}I were produced.

A key architect of the programme was Dr. V. K. Iya, often regarded as the "Pitamah of Isotopes," who helped build the scientific and institutional foundations for large-scale radioisotope production and applications in medicine, industry, and agriculture. The commissioning of the 40 MW CIRUS reactor in 1960 significantly expanded production capacity, supported by new radiological laboratories at Trombay. In the early 1970s, a modern radiological laboratories complex with shielded hot cells became operational, enabling larger-scale production of radiochemicals and radiopharmaceuticals.

Further growth came with the commissioning of the 100 MW DHRUVA reactor in 1985, followed by the creation of the Board of Radiation and Isotope Technology (BRIT) in 1989, which strengthened nationwide distribution of radioisotopes and radiation technologies.

3 Current Infrastructure for Medical Radioisotope Production in DAE

Medical radioisotopes in India are produced through both research reactors and particle accelerators operated under DAE. Reactor-based production remains the primary source of several therapeutic radioisotopes. Research reactors such as DHRUVA and APSARA-U at the BARC support the production of medically important isotopes including ^{64}Cu, ^{99}Mo, ^{131}I, ^{153}Sm, ^{177}Lu and ^{192}Ir.

A major recent addition to this infrastructure is the Fission Molybdenum Plant (FMP) at Trombay, established for large-scale production of fission-derived ^{99}Mo. The facility is expected to ensure a reliable domestic supply of ^{99}Mo for the preparation of ^{99}Mo/^{99m}Tc generators, which form the backbone of diagnostic nuclear medicine worldwide and strengthen India's self-reliance in the supply of critical medical radioisotopes.

Accelerator-based production complements reactor technologies, particularly for positron-emitting radioisotopes used in PET imaging. India's first medical cyclotron facility at the Radiation Medicine Centre (RMC), BARC, commissioned in 2002, routinely produces ^{18}F for PET radiopharmaceuticals such as [1^{8}F]FDG. In addition, the 30 meV medical cyclotron facility at the Variable Energy Cyclotron Centre (VECC), Kolkata, has expanded national capabilities for accelerator-based radioisotope production and supports research and development of novel radiopharmaceuticals.

4 Radiochemical Processing and Radiopharmaceutical Development

The production of clinical-grade radioisotopes requires sophisticated radiochemical separation and purification techniques performed in shielded hot-cell laboratories. Methods such as solvent extraction, ion-exchange chromatography, extraction chromatography and electrochemical separation are routinely employed. Following purification, radioisotopes are incorporated into target-specific radiopharmaceutical formulations. These preparations undergo rigorous quality control testing to ensure radionuclidic purity, radiochemical purity, sterility and apyrogenicity prior to clinical use. DAE has also made significant progress in the indigenous synthesis of targeting molecules and freeze-dried (lyophilized) kit formulations, allowing convenient preparation of radiopharmaceuticals in nuclear medicine departments. The major radioisotopes supplied from DAE are summarized in Tables 1 and 2.

Table 1. Summary of major radionuclides produced and deployed from BARC/BRIT

S. No.	Radioisotope	Nuclear reaction	Target	Radiochemical form	Major applications
1	^{131}I	(n,β^-)	$^{130}TeO_2$	[^{131}I]NaI	Treatment of thyroid disorders and neuroendocrine tumors (NETs)

(continued)

Table 1. (*continued*)

S. No.	Radioisotope	Nuclear reaction	Target	Radiochemical form	Major applications
2	^{177}Lu	(n,γ)	$^{176}Lu_2O_3$	$[^{177}Lu]LuCl_3$	Treatment of NETs, prostate cancer and palliative care
3	^{99}Mo	(n,γ)	$^{98}MoO_3$	$[^{99}Mo]Na_2MoO_4$	$^{99}Mo/^{99m}Tc$ generator
4	^{153}Sm	(n,γ)	$^{152}Sm_2O_3$	$[^{153}Sm]SmCl_3$	Palliative care
5	^{125}I	(n,EC)	^{124}Xe gas	$[^{125}I]NaI$	Brachytherapy

Table 2. Summary of major radionuclides and radiopharmaceuticals produced and deployed from VECC/BRIT

S. No.	Radioisotope	Half-life	Target	Nuclear reaction	Proton energy (MeV)	Radiopharmaceuticals
1	^{18}F	110 h	$H_2{}^{18}O$ (liquid)	(p,n)	18	$[^{18}F]FDG$, $[^{18}F]NaF$
2	^{201}Tl	73 h	^{203}Tl (solid)	(p,3n)	28	$[^{201}Tl]TlCl$
3	^{68}Ga	68 min	^{68}Zn (solid)	(p,n)	15	$[^{68}Ga]Ga$-PSMA-11, $[^{68}Ga]Ga$-DOTATATE

5 Medical Radioisotopes from "Wealth-From-Waste"

An important and unique aspect of India's radioisotope programme has been the recovery of useful radioisotopes from high-level nuclear waste generated during spent fuel reprocessing. This approach exemplifies the concept of "wealth from waste," where valuable radioisotopes are recovered and utilized for societal benefits. Radioisotopes such as ^{90}Sr for developing $^{90}Sr/^{90}Y$ generator for medical applications, ^{106}Ru for ophthalmic plaque therapy, and ^{137}Cs sources for irradiation technologies have been recovered by reprocessing waste streams. These efforts not only contribute to efficient resource utilization but also expand the availability of medically and industrially important radioisotopes.

6 Emerging Developments in Radioisotope Production

6.1 Lutetium-177 Production and Self-reliance in Radionuclide Therapy

BARC has established large-scale production of carrier-added (CA) ^{177}Lu, a key radioisotope used in therapy of neuroendocrine and prostate cancers using radiopharmaceuticals such as $[^{177}Lu]Lu$-DOTATATE and $[^{177}Lu]Lu$-PSMA-617, respectively.

Advances in target fabrication, irradiation, and processing have enabled India to achieve near self-reliance across the full ^{177}Lu value chain, from production to clinical use.

High-specific-activity no-carrier-added (NCA) ^{177}Lu is produced via the ^{176}Yb(n,γ)^{177}Yb $\rightarrow$ ^{177}Lu route. Indigenous separation methods include electrochemical processing at BARC and extraction chromatography at BRIT [2, 3].

6.2 Copper-64: Reactor Production

BARC reported the world's first production of clinical-grade ^{64}Cu using research reactors through innovative radiochemical processing of irradiated targets [4]. Reactor-produced [^{64}Cu]$CuCl_2$ has since been evaluated in PET imaging of cancer patients at several reputed hospitals in the country.

6.3 Targeted Alpha Therapy with Astatine-211

BARC has also demonstrated production of ^{211}At, an important alpha-emitting therapeutic radioisotope, using the K-130 cyclotron at VECC through the ^{209}Bi(α,2n)^{211}At reaction [5]. Ongoing efforts focus on developing ^{211}At-based targeted alpha therapies.

Additionally, extensive research is being conducted at BARC on the production, radiochemical separation, and purification of several other emerging reactor- and accelerator-produced radioisotopes, including, $^{43/44}$Sc, ^{52}Mn, ^{124}I, $^{132/135}$La, ^{161}Tb, ^{166}Ho, ^{169}Er, ^{186}Re, etc., for use in preparation of next-generation radiopharmaceuticals.

7 Development of Next-Generation Radiopharmaceuticals

Alongside large-scale radioisotope production, India has made major advances in next-generation radiopharmaceuticals through innovations in ligand design, radiolabeling chemistry, and therapeutic delivery systems. These efforts span molecular targeting agents, particulate systems, and nanomaterial-based platforms for theranostic applications.

7.1 Targeting Ligands for Molecular Imaging and Therapy

Indigenous synthesis of clinically important targeting ligands has strengthened domestic radiopharmaceutical capability [6, 7]. Examples include PSMA-617 for prostate-specific membrane antigen targeting in prostate cancer, and HYNIC-TOC for somatostatin receptor imaging in neuroendocrine tumors. Additionally, BARC has developed robust cold-kits (lyophilized ready-to-label formulations) for routine hospital radiopharmacy use, enabling rapid, on-site preparation of clinically relevant agents with improved shelf-life, quality assurance, and cost-effectiveness. These advances have substantially reinforced India's self-reliant innovation pipeline, accelerating the growth of personalized nuclear medicine and next-generation theranostic applications.

7.2 Radiolabeled Microparticles for Radiation Synovectomy

Particulate radiopharmaceuticals have been developed for localized therapy. Engineered hydroxyapatite microparticles labeled with ^{90}Y or ^{177}Lu have been evaluated for radiation synovectomy in rheumatoid arthritis [8]. Intra-articular administration enables targeted irradiation of inflamed synovial tissue, reducing pain and improving joint function.

7.3 Intrinsically Radiolabeled Nanoparticles for Cancer Therapy

Nanotechnology has facilitated the development of intrinsically radiolabeled nanoparticles, wherein the radioisotope is incorporated within the particle matrix, thereby enhancing in vivo radiochemical stability and functional integration. Representative systems such as [^{177}Lu]Lu_2O_3, [^{188}Re]ReO_x, and [^{169}Yb]Yb_2O_3 nanoparticles, frequently surface-engineered with biocompatible protein coatings, have demonstrated considerable potential for multimodal imaging and synergistic therapeutic applications in preclinical cancer models [9]. Importantly, [^{177}Lu]Lu_2O_3 nanoparticles have progressed to clinical evaluation for the treatment of locally advanced breast cancer.

7.4 BhabhaSpheres for Liver Cancer Therapy

A major indigenous innovation is ^{90}Y glass microspheres (BhabhaSpheres), developed at Bhabha Atomic Research Centre for treatment of unresectable liver cancer [10]. Produced by neutron activation of yttrium-containing glass particles, these microspheres are administered via the hepatic artery, where they lodge in tumor vasculature and deliver localized β^--radiation. As a cost-effective domestic alternative to imported products, BhabhaSpheres can significantly improve access to advanced liver cancer therapy in India.

Collectively, these developments highlight India's growing capability to integrate radioisotope production, ligand chemistry, and advanced therapeutic platforms, supporting the expanding field of theranostics and personalized nuclear medicine.

8 Future Facilities and Next-Generation Technologies

The Indian radioisotope programme is poised for significant expansion with several upcoming facilities. The Isotope Production Reactor (IPR) and Isotope Processing Laboratory (IPL) being established at BARC Facility, Visakhapatnam, will significantly enhance production of medical radioisotopes such as ^{99}Mo, ^{131}I, ^{177}Lu, ^{188}W and ^{192}Ir. Together with the recently commissioned FMP at Trombay, these initiatives are expected to ensure reliable domestic availability of key radioisotopes required for nuclear medicine.

Accelerator-based technologies are also opening new avenues. The Radioactive Ion Beam (RIB) facility at VECC, Kolkata aims to produce novel radioisotopes with high specific activity. Complementing this effort is the Stable and Unstable Isotopic Beam for Heavy Ion Research (SUBHIR) facility under development at BARC Facility, Vishakhapatnam, which will enable production of emerging medical radioisotopes.

9 Conclusions

Radioisotopes have become indispensable tools in modern healthcare, enabling highly sensitive diagnostic imaging and targeted radionuclide therapy. Over several decades, India has developed a strong and largely self-reliant radioisotope programme supported by advanced reactor and accelerator infrastructure, radiochemical expertise and clinical collaborations. With continuing advances in theranostic radioisotopes, targeted alpha therapy and next-generation production facilities, India is well positioned to further expand its contributions to nuclear medicine and healthcare.

Acknowledgments. The author gratefully acknowledges that the work summarized in this chapter reflects not only his own contributions but also the collective efforts of many colleagues and collaborators over several decades. He sincerely thanks his colleagues in the Radiopharmaceuticals Division, BARC, as well as scientists and engineers from other Divisions of BARC and various units of the DAE. The author also respectfully recognizes the contributions of successive generations of scientists and technologists whose dedication has built and sustained India's radioisotope programme.

References

1. Bodei, L., Herrmann, K., Schöder, H., Scott, A.M., Lewis, J.S.: Radiotheranostics in oncology: current challenges and emerging opportunities. Nat. Rev. Clin. Oncol. **19**(8), 534–550 (2022)
2. Patra, S., Chakravarty, R., Singh, K., Vimalnath, K.V., Chakraborty, S.: Electrochemical separation and purification of no-carrier-added ^{177}Lu for radiopharmaceutical preparation: translation from bench to bed. Chem. Eng. J. Adv. **14**, 100444 (2023)
3. Kumar, D., et al.: A PLC based semi-automated extraction chromatographic separation system for the isolation of medical grade no-carrier-added lutetium-177 for targeted cancer therapy. React. Chem. Eng. **10**, 1569–1576 (2025)
4. Chakravarty, R., Chakraborty, S., Dash, A.: $^{64}Cu^{2+}$ ions as PET probe: a new paradigm in molecular imaging of cancer. Mol. Pharm. **13**, 3601–3612 (2016)
5. Ghosh, S., Banerjee, D., Guleria, A., Chakravarty, R.: Production, purification and formulation of nanoradiopharmaceutical with ^{211}At: an emerging candidate for targeted alpha therapy. Nucl. Med. Biol. **138–139**, 108947 (2024)
6. Ajish Kumar, K.S., Mathur, A.: A convenient total synthesis of PSMA-617: a prostate specific membrane antigen (PSMA) ligand for prostate cancer endotherapeutic applications. Eur. J. Med. Chem. Rep. **6** 100084 (2022)
7. Kar, S., et al.: Convenient one-pot synthesis of 1-(4-(4-(2-[^{18}F]fluoroethoxy)phenyl)piperazin-1-yl)ethanone ([^{18}F]FEt-PPZ) for imaging tumors expressing sigma-1 receptors. RSC Adv. **15**(29), 23943–23953 (2025)
8. Patra, S., et al.: Radiolabeled nanoporous hydroxyapatite microspheres: an advanced material for potential use in radiation synovectomy. Mat Chem Phys **295**, 127115 (2023)
9. Ghosh, S., et al.: Protein-functionalized and intrinsically radiolabeled [^{188}Re]ReO_x nanoparticles: advancing cancer therapy through concurrent radio-photothermal effects. Eur. J. Nucl. Med. Mol. Imaging **52**(7), 2628–2644 (2025)
10. Vimalnath, K.V., et al.: [^{90}Y]Yttria Alumino silicate glass microspheres: a biosimilar formulation to 'TheraSphere' for cost-effective treatment of liver cancer. Cancer Biother. Radiopharm. **39**(1), 82–91 (2024)

Megascience Frontiers

Mega Science Frontier in Nuclear and High-Energy Physics: Present and Future

Amol Dighe(✉)

Department of Theoretical Physics, Tata Institute of Fundamental Research, Homi Bhabha Road, Colaba, Mumbai 400005, India
amol@theory.tifr.res.in

Abstract. Mega science projects (MSPs) are large collaborative projects that address major scientific problems, involving multiple institutions and requiring large quanta of funding. This talk will outline the Mega Science Vision—2035 reports in Nuclear Physics and High Energy Physics, which have recently been published by the Office of the Principal Scientific Advisor, Government of India. Some suggestions for the way forward will be discussed.

Keywords: Mega Science · Nuclear Physics · High Energy Physics

1 Introduction

The Mega Science Vision-2035 (MSV-2035) exercise was initiated by the Office of the Principal Scientific Advisor (PSA) to the Government of India in December 2020. The process was started in six disciplines: (i) nuclear and plasma physics, (ii) high-energy and astroparticle physics, (iii) accelerator S&T and applications, (iv) astronomy and astrophysics, (v) climate research, and (vi) ecology and environmental science. While the first four reports have been completed and published at this time, work on the last two is ongoing. This talk will focus on the first two of these reports, denoted here as MSV-NP [1] and MSV-HEP [2], respectively. The next two reports [3, 4] will be discussed as a part of the talks by Prof. B. C. Joshi and Prof. K. K. Pant in this DAE Conclave. The emphasis will be on informing the reader on what these reports contain, what they recommend, how they are likely to be used, and how the scientific community can benefit from them.

The MSV-2035 exercise has been an inclusive one, where the scientific aims and ambitions of the Indian scientific community in respective disciplines have been reflected. For this purpose, in each subject area, a Working Group (WG) was formed whose combined membership had a broad knowledge of the pressing scientific questions in that area, and activities being pursued by Indian scientists. The WG had representatives from the universities, institutes, other academic institutions, experimental collaborations, as well as funding agencies. The information collected by the WG members went to the Drafting Group (DG), a subset of the WG who were involved in actual writing of the Report. After a convergence on the content of the document within the WG, inputs were solicited from all cross sections of the subject community, and their suggestions were

G. B. Mohanty and V. S. Prabhudesai (Eds.): DAEC 2026, SPPHY 368, pp. 153–158, 2026.
https://doi.org/10.1007/978-981-92-2948-2_19

incorporated. The draft then was vetted by a set of national and international experts, and finally by the Office of the PSA.

The discussions within the WG focussed on the following issues: (i) important science questions that need to be pursued collectively, (ii) questions that can be addressed practically in the next 10–15 years, given the experience, expertise and resources available in the country, (iii) building up the infrastructure/ecosystem for future, (iv) facilitating the inception of new mega science projects (MSPs), (v) possible spinoffs and industry involvement, (vi) human resource and funding needs. These discussions have led to some introspection about what we have learnt over the years, what expertise we have gained, and what scale of projects we can implement. The reports will also serve as a repository of information about MSPs that Indian scientists are currently involved in.

2 Nuclear and Plasma Physics (MSV-NP)

This can be divided into roughly three categories: (i) quantum chromodynamics (QCD), (ii) nuclear structure and dynamics, (iii) plasma physics.

In the area of quantum chromodynamics, some of the important open questions are: What is the phase structure of QCD matter? How do the strong interactions amongst quarks and gluons inside the nucleons result in confinement and collectively result in their properties such as mass and spin? How does a nucleus look in terms of its constituents? To be able to address these questions, the following recommendations have been made: (i) Participation in heavy-ion programs at the international facilities such as LHC, RHIC and FAIR should be continued, since mapping the QCD phase diagram would be possible only when the data from these are taken together. (ii) The CBM experiment, which is under construction at FAIR, should be the focus for the high-energy nuclear collisions in near future. (iii) Participation in the upcoming Electron-Ion Collider experiments would be desirable.

As far as the nuclear structure and dynamics is concerned, some of the pressing questions are: How does subatomic matter organize itself (nuclear shapes, sizes, spins)? How does the equation of state of nuclear matter decide properties of neutron star? What will neutron star merger tell us about synthesis of heavy elements? What is the origin of neutrino mass and are they their own antiparticles? Are there neutrinos species beyond the standard model? What is dark matter and what are its constituents? To address these, the following are recommended: (i) development of new accelerator facilities within India for radioactive-ion beams (RIBs), (ii) underground facilities for low-energy nuclear physics programs, (iii) a strong participation in the experiments at FAIR and other international nuclear physics facilities for RIBs and photon beams, (iv) state-of-the-art detector systems at the existing accelerator facilities, (v) a consortium to facilitate the usage of some of the international facilities by the low-energy nuclear physics groups in India

Plasma physics needs to address questions such as: What are the physico-chemical processes involved in plasma-liquid/solid interactions? Can collective behaviour of ultracold plasma be exploited for quantum computing? How can we understand astro-plasmas using multiscale collision-less reconnections? This requires (i) continued participation in the ITER experiment, (ii) setting up plasma-based systems for generating an ultra-cold plasma trap for quantum computing, (iii) an astroplasma simulator at lab scale, (iv) building a moderate-sized spherical tokamak that can be used for the production of radioactive sources for medical purposes, production of tritium for conventional fusion reactors, or as the core of a fission-fusion hybrid reactor, (v) setting up a high-energy (MJ), high power (TW) pulsed-power facility for producing intense plasma.

3 High-Energy and Astroparticle Physics (MSV-HEP)

The pathways for the future in the area of high-energy and astroparticle physics may be represented by Fig. 1. Note that detector R&D and high-performance computing are the foundations on which the success of MSPs rests.

In order to retain and enhance the leadership of India in HEP MSPs, the following actions should be given top priorities: (i) *Collider experiments*: Support to the current MSPs, CMS and Belle II, should be continued. In addition, planning and R&D efforts for experiments at future colliders such as FCC-ee and/or a muon collider should start. Instrumentation R&D for new technologies for future detectors (e.g. quantum sensors) should be encouraged. (ii) *Underground experiments*: Efforts to establish an underground observatory in the country for experiments in HEP, nuclear physics and astrophysics should be reinvigorated. A dark matter search experiment—such as InDEX at Jaduguda—that uses emerging technologies could become a flagship experiment here. (iii) *Neutrino experiments:* Participation in DUNE, especially if there is possibility of building an important component of detector, is a good opportunity. (iv) *Cosmic ray experiments*: Continued support for GRAPES-3 and MACE, and the neutrino scattering experiment ICNSE is recommended. Stereo MACE should be aggressively pursued. Participation in CTA is strongly recommended, given its synergy with MACE.

From the perspective of long-term future, it is crucial to give a push to develop indigenous projects that may be currently small-scale but may grow into MSPs. Some examples could be the dark matter search experiment InDEX mentioned above, neutrinoless double beta decay experiments such as TinTin, observation of low-energy solar neutrinos using deuterated liquid scintillator (all of which would need an underground science facility to reach their full potential), reactor monitoring and sterile neutrino search experiment such as ISMRAN, or satellite-based experiments exploring the Universe in medium-energy gamma rays.

There are leading international mega experiments such as Hyper-Kamiokande and IceCube-Gen2 in which Indian scientists have been participating in an individual capacity. If these are to be treated like a mega science activity in a country, the existence of a critical number of interested and dedicated scientists needs to be demonstrated.

4 Infrastructure-Building for Future Mega Science Projects

In addition to participating in the international MSPs, which helps in building expertise in the country and for knowledge/technology transfer, for an emerging economy such as India, it is also important to be able to host MSPs in the country, where international community can participate. Apart from being a prestigious activity that showcases the

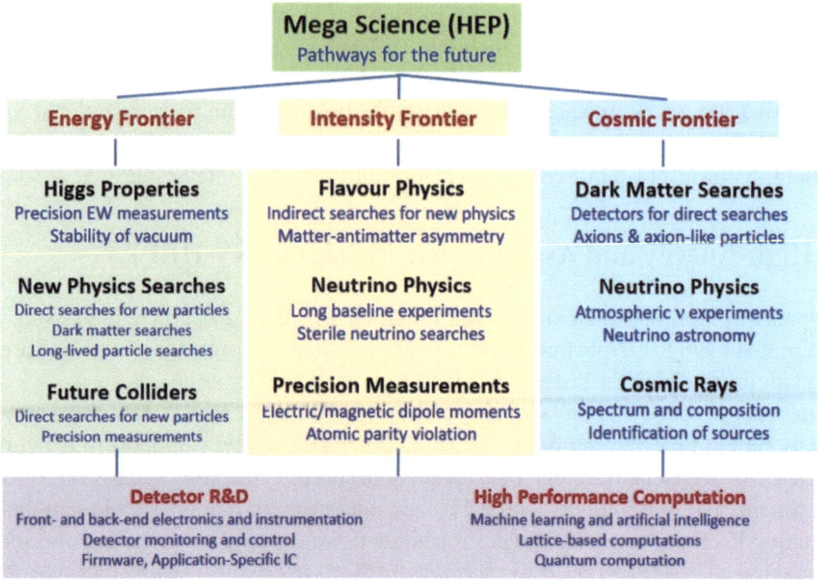

Fig. 1 Pathways for the future in high-energy and astroparticle physics (reproduced from MSV-HEP).

scientific strength and maturity of the country, it would also bring Indian scientists, students and population closer to scientific action on the world stage, provide a wider exposure, and help in training future scientists.

MSPs cannot start spontaneously. They need a wide base of scientific activity, expertise, and infrastructure in the country. Some of the strong recommendations in order to develop such an infrastructure are: (i) establishment of a National Detector Development and Training Centre (NDDTC) which can host common facilities for detector development, state-of-the-art laboratories and/or workshops for instrumentation, testing and computation, available for users all over the country, (ii) an underground science facility (USF) for projects that look for rare events and hence need a location that is relatively free from the cosmic ray background, (iii) research reactors, photon beam and ion beam facilities, (iv) data centres distributed over the country for high-performance computation.

The process of initiation of new mega science projects may be streamlined by a standardized procedure of Community Planning Exercise, a single-window system – a Mega Science Coordination Unit (MSCU) – and a separate line in the budgets for mega science activities. Taking a long-term view, experiments that are not at the level of a MSP at the moment but have the potential to develop into MSPs over a longer duration, should be considered for funding under these activities.

Fostering industry participation in MSP activities is important for these activities to be scalable. Some initiatives on this front may be (i) allowing industry to directly be part of the MSPs, (ii) granting — with oversight — the involved industry the rights to use the results of the research at the project-end, (iii) developing avenues for students, postdocs and staff to gain experience by working in an industrial setting, (iv) providing support for high-risk/high-gain research proposals aimed at developing a generic technology.

5 Next Steps

The MSV documents are the visions of the scientific communities. They are not approved by the funding agencies or the government. However, this is the first time such extensive formal documents have been prepared in the country. It is expected that these will be referred to by the concerned government agencies to get an idea of what the scientific community desires, and it is hoped that some of the suggestions will be deliberated on and implemented. However, initiatives for action need to be taken by scientists.

Those with ambitious plans of establishing a MSP should come together with a proposal, utilizing some of the recommendations as evidence of community support for the science case. They should try to convince the broader community of the scientific worth and feasibility of the project, carry out a Community Planning Exercise, and write a White Paper to be submitted to the funding agencies or the proposed MSCU.

Some closely related activities have already been taking place, in parallel with or as follow-ups of the MSV exercise. The DAE Amrit Kaal Vision Documents for R&D Units, Public Sector Enterprises, and Aided Institutions have recently been published [5]. The priorities mentioned therein have a large overlap with those in the MSV-2035, and many aspects are complementary. An internal meeting on "Mega Science in nuclear physics, high energy physics and astrophysics" was held in TIFR, Mumbai, in October 2025, to seek commonalities in interests and instrumentation. TIFR scientists have proposed to set up a detector R&D center for next-generation technologies, and to work towards an Indian facility for matter-antimatter research. A special session on "Science Opportunities at an Underground Science Facility" was held in the Annual Meeting of Indian Academy of Sciences, Bengaluru, in November 2025. Indian Nuclear Physics Association (INPA) is in the process of being formed. More such follow-ups are expected and should be encouraged.

The areas of nuclear and high energy physics have synergies among themselves (areas of dark matter, neutrinos, QCD, heavy-ion collisions, quark-gluon plasma, detector development) as well as with the accelerator R&D and applications (particle colliders, accelerators, cyclotrons, pelletron-LINAC, high-intensity ion beams, radioactive ion beams) and with astronomy and astrophysics (astroparticle physics, multi-messenger astronomy, gravitational waves). The long-term visions of Indian scientists in these four related fields, where DAE has a strong presence, should therefore be developed and implemented together. सह वीर्यं करवामहै।

Acknowledgments. The author would like to acknowledge the working group and drafting group members of MSV-NP and MSV-HEP for all their efforts in preparing the documents, and the funding from DAE under Project Identification Number RTI 4012.

References

1. Mega Science Vision—2035 Nuclear Physics, ISBN:978–81–991742-7-6, (2023). Available at https://www.psa.gov.in/mega-science-vision

2. Mega Science Vision—2035 (High Energy Physics), ISBN:978–81–991742-6-9, (2025). Available at https://www.psa.gov.in/mega-science-vision
3. Mega Science Vision—2035 (Accelerator S&T and Applications), ISBN:978–81–991742-9-0, (2024). Available at https://www.psa.gov.in/mega-science-vision
4. Mega Science Vision—2035 (Astronomy and Astrophysics), ISBN:978–81–991742-4-5, (2024). Available at https://www.psa.gov.in/mega-science-vision
5. DAE Amrit Kaal Vision Documents: Available at the DAE Digital Library. https://digilib.dae.gov.in/books/10736

Sensing Gravitational Waves with the Galactic Clocks

Bhal Chandra Joshi[1,2](✉)

[1] National Centre for Radio Astrophysics, Tata Institute of Fundamental Research, P Bag No 3, Ganeshkhind, Pune 411007, Maharashtra, India
bcj@ncra.tifr.res.in

[2] Department of Physics, Indian Institute of Technology, Roorkee 247667, Uttarakhand, India

Abstract. Gravitational waves (GWs), predicted by Albert Einstein a century earlier, were finally discovered 10 years back by terrestrial detectors. Their ultra-long wavelength counterparts can be sensed by a Galactic scale detector consisting of an ensemble of rotating massive compact neutron stars called radio pulsars. Compelling evidence for these GWs was found in June 2023 by Indian Pulsar Timing Array (InPTA), alongwith other international experiments, using a sample of pulsars observed with the Udhagmandalam Radio Telescope (ORT) and the upgraded Giant Meterwave Radio Telescope (uGMRT), operated by the Department of Atomic Energy(DAE). A general introduction to the subject, the current status and the future landscape is outlined in this article.

Keywords: Gravitational waves · pulsars · Galaxies · inter-stellar medium

1 Introduction

We learn about the evolution and physical processes in our Universe through four messengers. While the electromagnetic waves (EM), cosmic rays, and neutrinos were discovered earlier, the latest of these messengers, gravitational waves (GWs), were discovered in 2015 by terrestrial detectors [1]. This path breaking discovery found GWs at short wavelengths or frequencies of about few Hertz. The quest for their ultra-long wavelength counterparts (nanoHertz GWs : nHzGWs) is currently ongoing by six international pulsar timing array (PTA) experiments. These include our effort in India called the Indo-Japanese pulsar timing array experiment [InPTA: 2–4], which uses the DAE's two premier radio astronomy observatories, *viz.*, the upgraded Giant Meterwave Radio Telescope (uGMRT) and the Udhagmandalam Radio Telescope (ORT). The InPTA pools its data with the European PTA [EPTA: 5, 6], the North American Nanohertz Observatory for Gravitational waves [NANOGrav: 7, 8], the Australia-based Parkes PTA [PPTA: 9, 10], the Chinese PTA [CPTA: 11] and MeerKAT PTA [MPTA: 12], under the auspices of the International Pulsar Timing Array consortium (IPTA).

G. B. Mohanty and V. S. Prabhudesai (Eds.): DAEC 2026, SPPHY 368, pp. 159–167, 2026.
https://doi.org/10.1007/978-981-92-2948-2_20

The current status of these experiments in sensing GWs and the future roadmap is presented in this brief review. GWs are introduced in Section 2 highlighting their role in astrophysics. Next, the sources of GWs are covered with the respective detectors in Section 3 with a brief explanation of the operating principle of PTAs. The InPTA experiment is described in Section 4. The current status of PTA science is presented in Sections 5 highlighting the current challenges. The paper concludes by outlining future roadmap in Section 6.

2 Gravitational Waves

GWs are propagating perturbation of space and time, analogous to electromagnetic (EM) waves, and are a natural consequence of description of gravity as a geometrical construct by Einstein [13]. GWs are emitted due to dynamics of asymmetric mass distributions, such as two orbiting supermassive black holes (MBH) or in explosions of massive stars to cosmological phase transitions. For a binary source, the frequency of GWs is twice its orbital period. For the widely accepted Einstein's General Theory of Relativity, the waves are transverse, quadrupolar and with two polarizations, although this could be different for alternative gravity theories [14].

GWs are expected to be quadrupolar and couple weakly to matter compared to EM waves. While this makes it challenging to detect these waves, these waves propagate across the Universe without scattering and absorption penetrating over distances where astrophysical information is lost for other messengers in astronomy. Besides, we are probing the most massive and compact objects in the universe, which provides tests of theories of relativity in the strongest regime and sheds light on fundamental constituents of matter and nuclear physics. Therefore, GW astronomy, a very new and evolving field, is expected to have a significant impact on fundamental physics.

While the highest frequency GWs are emitted by binary systems with compact stellar mass objects, such as neutron stars (NS) and black holes (BH), the GW frequency reduces progressively with increasingly massive objects such as MBH binaries or GWs of cosmological origins. Similar to EM waves, the detection methods for GWs for different frequencies require different type of detectors ranging from terrestrial to Galactic detectors.

3 Detectors

GWs from different sources are sensed using detectors corresponding to GW wavelength. Earth based detectors can be used to sense GWs emitted during coalescence of compact stellar mass binary sources, such as NS or BH binaries, as their wavelength are about 10000 km. On the other hand, the origin of ultra-long wavelength GWs with wavelengths of a parsec (10 trillion km) is very different. Often, MBHs (typically with billion solar mass) at the center of the two merging galaxies are driven towards each other by the friction provided by the gas and stars in the galaxy eventually forming a MBH binary. This binary has a GW radiation dominated in-spiral ending with eventual coalescence [15]. During the gradual in-spiral, continuous GWs (CGWs) are emitted, while the coalescence of two MBH is observed as GW burst. While the emssion from individual MBH binary is expected to be too weak to be detected, an incoherent sum of CGWs over all such binaries at different frequencies is expected to produce an isotropic stochastic GW background (SGWB) with a characteristic power law spectrum with a index of 2/3 [Fig. 4: 15]. A Galactic detector is required to sense these waves. The principle of operation of relevant detectors for both these wavebands are discussed below.

3.1 Interferometer

A passing GW distorts the space-time. Therefore, GW can be detected by measuring the variation of proper distance between two points in space-time. Terrestrial detectors like Advanced Laser Interferometer Gravitational-Wave Observatory [aLIGO: 16] employ a laser interferometer. The arms of interferometer are adjusted to get a "dark" fringe. When a GWs passes the interferometer, the stretching/shrinking of arms translates to variations in brightness for this fringe, which constitutes a measurement of GW strain. With a laser beam reflected 300 times over an arm length of 4 km, aLIGO is sensitive to shorter wavelength GWs or frequencies in the range of 10 - 500 Hz. An Indian component of aLIGO is under construction at Hingoli, India funded by DAE and DST, which will aid in better localization of GW sources in the sky [17]. More advanced terrestrial detectors are planned in future [18–20].

3.2 Pulsar Timing Array

The ultra-long wavelength GWs require a Galactic size detector. Here, the variation of proper distance between two points in space-time is sensed by measuring variations in propagation delay of pulsed EM signals from compact massive stars called radio pulsars, which are located at a typical distance of kiloparsec.

Radio pulsars are compact massive highly magnetized rotating neutron stars, which emit clock like radio pulses. Millisecond pulsars (MSPs), with short periods and smallest magnetic fields, have a very stable rotation and hence provide the most precise clock ticks in the Universe allowing measurements of delays as small as few nanoseconds. As the pulse travels from the pulsar to our radio telescope, the distortion in space-time changes the arrival time of the pulsar and hence the apparent frequency of this clock. A passing gravitation wave therefore produces variations in tone of the pulsar, so in a sense we use radio waves to listen to these GWs. In practice, this is done by subtracting the observed time of arrival from that predicted by a model of rotation of the star, delays in propagation and relativistic and binary effect. In absence, of GWs, the result of this subtraction, called timing residuals, are close to zero and normally distributed, but would show a sinusoidal signal if a CGW is present.

Fig. 1. Skill development schools funded by DAE and organised by InPTA collaboration

The variations in the radio pulse arrival can be due to a small random wander in the pulsar rotation, called timing noise or due to variations in refraction, dispersion and scattering by the free electrons in the ionised inter-stellar medium (IISM). The former "noise" is independent of observing radio frequency (achromatic), while the latter (chromatic) is more prominent at radio frequencies below 1 GHz. These variations are of the order of μs compared to nanosecond level variations expected from GWs. Therefore, the measurement of the tiny delays due to GWs is challenging.

This challenge can be overcome by observing a pair of widely separated pulsars. While the achromatic timing noise and chromatic propagation noise is uncorrelated between the pair, the GW imprint is correlated across the pair as a function of the angular separation of pulsars [Fig. 2: 21]. This correlation is called Hellings-Down overlap reduction function [22] and is like the response function of a PTA formed from a large ensemble of MSPs.

The characteristic spectrum of of SGWB covers frequencies in the nanoHertz range [15]. To extract this spectrum from the data, a PTA requires to observe pulsars over a time baseline of several years (inverse of required frequency). Therefore, a PTA experiment observes regularly a large collection of MSPs with a large radio telescope to time them and extract the above correlation and spectrum from these observations.

4 Indo-Japanese Pulsar Timing Array (InPTA)

An Indian PTA experiment for discovering GWs was originally visualized by Prof Swarup in mid-1990s even as he was building the GMRT. It took almost two decades for us to tune the GMRT for this purpose. Initial efforts started with the ORT in 2012 followed by observations with the GMRT and uGMRT between 2015–2018. A multi-institution collaboration for running this experiment was established in 2018[2]. The required person power (under-graduate, graduate and doctoral students)was trained from early days itself through several radio astronomy schools, student weeks, and busy weeks funded by DAE at Pune, Chennai, Mumbai, Hyderabad, Ooty, and Roorkee by the faculty involved at these institutions (Figure 1). This approach of training person-power while carrying out cutting edge radio astronomy observations led to rapid increase in members to about 40 from 11 institutions in India and Japan alongwith a rapid science exploitation of the data by this person power. Thus, InPTA has been a good skill development program in addition to being a major science program. The InPTA became a full member of IPTA in March 2021 [3].

The InPTA observes about 30 pulsars every 14 days collecting about 50 TB of data over 24 observations every year. The InPTA uses the unique subarray capability of the uGMRT to form two co-located 75-m telescopes for concurrent 300–500 and 1260–1460 MHz radio observations. This observing mode makes the InPTA data set best suited amongst all PTA experiments for high precision DM measurements and for characterization of IISM noise. We have achieved unprecedented measurement of chromatic noise in the history of pulsar astronomy due to this unique capability. This makes the InPTA data very valuable till the SKA Observatory comes up in 2035. Additionally, the InPTA experiment provides a template for a PTA with the SKA Observatory [3].

The InPTA made its first data release (DR1) in 2022 with observations of 14 pulsar over 3.5 years[23]. This has now been followed by the second InPTA data release (DR2) with observations of 27 pulsars over 7 years[4]. DR2 provides the best estimates of chromatic dispersion noise in the world with unprecedented DM precisions of one part in 10^{-6} with the DM time-series clearly showing signatures of solar wind. In 2021, the InPTA complemented the 20 year long time baseline high radio frequency European EPTA data by combining the unique low radio frequency InPTA DR1 data. One of the immediate outcome was learning that high frequency data can be misleading due to leakage between achromatic and chromatic noise as the latter is not estimated well at high frequency observations[24]. Therefore, including 300–500 MHz InPTA data improved the noise modeling and the sensitivity of GW extraction.

The results of search for SGWB in the combined InPTA and EPTA data, announced in June 2023, showed for the first time a compelling evidence for nHzGWs opening up a window in GW spectrum complimentary to that discovered by aLIGO in 2015 [Fig. 5: 25]. This significant achievement was much like the discovery of radio emission in 1930s. This evidence was independently corroborated by CPTA, PPTA and NANOGrav experiment in coordinated announcement of this discovery in June 2023 [26–28]

While an astrophysical origin of SGWB is a natural explanation, there are several alternatives and exciting and exotic possibilities. These include different ways the SGWB can arise from a cosmological origin, such as inflation, phase transitions, and from topological defects as well as more exotic alternatives, such as invoking ultra-light dark matter or primordial black holes. The PTA experiments, including the InPTA, cannot discriminate between these different scenarios currently although this may be possible in the next two decades with the PTAs sampling longer wavelengths due to a progressively increasing time baseline. Combining the InPTA data, particularly DR2 and upcoming Decadel InPTA data release 3 (DR3), with other PTAs under the auspices of IPTA is another way to improve sensitivity to test this exciting GW astrophysics.

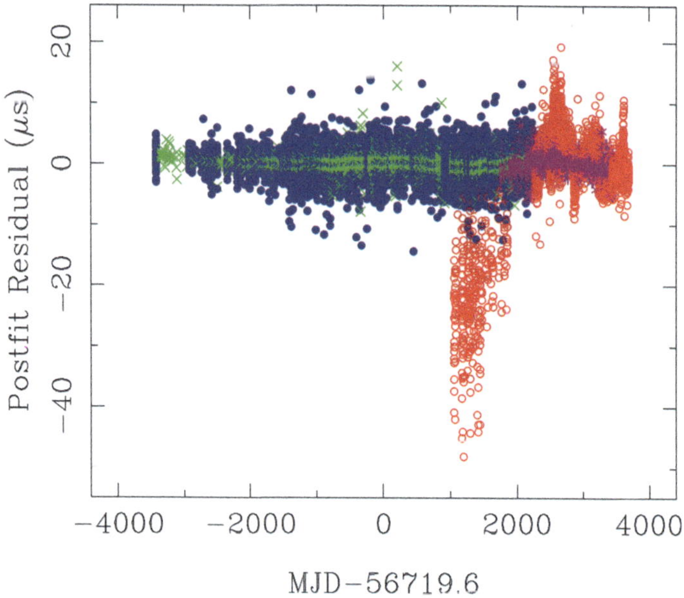

Fig. 2. Timing residuals for InPTA DR2 concurrent dual band data (300–500 MHz data is plotted in red colour) plotted against EPTA DR2, NANOGrav 15 year and PPTA DR3 data [6, 8, 10].

5 Current Status

The data from the six experiments - InPTA, EPTA, PPTA, NANOGrav, CPTA and MPTA, and two auxiliary telescopes - CHIME and LOFAR, are currently being combined under the auspices of IPTA as its third data release (IPTA DR3). This dataset with 131 pulsars forming more than 8500 pairs across the sky is the most sensitive dataset ever made with the expectation of a significant five sigma detection of the SGWB. In this combination, the InPTA DR2 will provide the low frequency coverage critical for accounting chromatic noise. A clear signature of large variations of chromatic noise is visible in the model subtracted pulse arrival time residuals for InPTA low frequency data (Figure 2) compared to significantly smaller imprint in the high frequency data of other PTAs. Thus, the InPTA DR2 adds a rich dimension to IPTA, particularly as direct dispersion noise measurements of InPTA can be used to model chromatic noise better independent of achromatic noise. This would play a major role in extracting GW signal in the upcoming IPTA DR3.

The robustness of customised noise models derived from InPTA DR2 data is evident from noise subtracted timing residuals, which show much less non-Gaussianity [Figs. A2, A3 and A4: 29]. Moreover, the posterior distributions for the amplitude and spectral slope for the achromatic and chromatic noise derived from InPTA DR2 data [Figs. 2, 3 and 4: 29] suggest little leakage into GW signal thereby improving sensitivity of the data set to GWs. The impact of solar wind corruption is seen in posteriors for estimated timing noise with and without data near solar conjunction for PSR J1022+1001 using InPTA DR2 data [Fig. C1: 29] and combined InPTA+LOFAR+EPTA data [Fig. 5: 30]. These results have motivated new solar wind models, such as periodic, Taylor series and short duration correlation solar models.

6 Future Roadmap

A search for GWs in the seven year InPTA data is currently ongoing using supercomputers at IIT Hyderabad, IIT Roorkee and IISER Bhopal. Likewise, a search for burst with memory is also being developed at IIT Roorkee. The InPTA team has already moved on to producing its decadel data release (InPTA DR3), which will have a ten year time baseline. Planned observations with the uGMRT are likely to continue for the next 5 years to increase the baseline.

The combination of InPTA DR2 data with the rest of the IPTA was recently completed by a team of 20 InPTA members and the single pulsar noise analysis is currently in progress using supercomputers at Physical Research laboratory, IIT Hyderabad and IIT Roorkee. This major international effort is likely to find in-controvertible five standard deviation detection of GWs in IPTA DR3 by the end of 2026, thus transforming ultra-long wavelength GW astronomy beyond detection to constraining GW astrophysics, such as discovering anisotropy in the SGWB.

A great leap in the ultra-low wavelength GW astronomy will happen with now under-construction SKA Observatory. This is the world's next big telescope jointly being built by consortium of 13 nations, where India is full member with the funding provided by DAE and DST. The SKA observatory is actually two telescopes, both intereferometers, much like uGMRT. The SKA-low, with 512 stations observing between 50−350 MHz, is located in Australia, whereas the SKA-mid, consisting of 197 dish antennas observing between 0.35−8 GHz is located in South Africa[1]. Both these sites were chosen because these sites have the lowest human made radio chatter or noise. Pulsar astronomy is a prime science case for the SKA Observatory[31], but the top priority in the pulsar science

[1] https://www.skao.int/en

with the SKA Observatory is the SKAO-PTA. Not only the SKA Observatory is likely to significantly increase the pulsar sample for the SKAO-PTA from the current 131 to about 300 MSPs, it will achieve better sensitivity to GWs using a concurrent multi-band monitoring like the InPTA with robust noise models in a 5 year campaign. With the 20 years of legacy data, this will transform the ultra-long wavelength GW astronomy from detection to to post-detection astrophysics era probably discriminating between astrophysical and cosmological origin of SGWB, uncovering anisotropy in this background and constraining Galaxy evolution history of the Universe[32]. Thus, sensing GWs with these facilities, funded by DAE, is likely to see an exciting decade to come.

Acknowledgment. BCJ acknowledges the support from Raja Ramanna Chair fellowship of the Department of Atomic Energy, Government of India (RRC Track I Grant 3/3401 Atomic Energy Research 00 004 Research and Development 27 02 31 1002//2/2023/RRC/R&D-II/13886 and 1002/2/2023/RRC/R&DII/14369).

References

1. Abbott, B.P., et al.: Observation of gravitational waves from a binary black hole merger. Phys. Rev. Lett. **116**, 061102 (2016)
2. Joshi, B.C., et al.: Precision pulsar timing with the ORT and the GMRT and its applications in pulsar astrophysics. J. Astrophys. Astron. **39**, 51 (2018)
3. Joshi, B.C., et al.: Nanohertz gravitational wave astronomy during SKA era: an InPTA perspective. J. Astrophys. Astron. **43**, 98 (2022)
4. Rana, P. et al.: The Indian Pulsar Timing Array data release 2: I. Dataset and timing analysis. Publ. Astron. Soc. Aust. **42**, e108 (2025)
5. Desvignes, G., et al.: High-precision timing of 42 millisecond pulsars with the European Pulsar Timing Array. Mon. Not. R. Astron. Soc. **458**, 3341–3380 (2016)
6. EPTA Collaboration et al.: The second data release from the European Pulsar Timing Array. I. The dataset and timing analysis. A&A **678**, A48 (2023)
7. Jenet, F.: The North American Nanohertz observatory for gravitational waves (2009). arXiv:0909.1058
8. Agazie, G., et al.: The NANOGrav 15 yr data set: observations and timing of 68 ms pulsars. ApJL **951**, L9 (2023)
9. Manchester, R.N., et al.: The Parkes Pulsar Timing Array project. Publ. Astron. Soc. Aust. **30**, e017 (2013)
10. Zic, A., et al.: The Parkes Pulsar Timing Array third data release. Publ. Astron. Soc. Aust. **40**, e049 (2023)
11. Xu, H., et al.: Searching for the Nano-hertz stochastic gravitational wave background with the Chinese pulsar timing array data release I. Res. Astron. Astrophys. **23**, 075024 (2023)
12. Miles, M.T., et al.: The MeerKAT Pulsar Timing Array: first data release. Mon. Not. R. Astron. Soc. **519**, 3976–3991 (2023)
13. Einstein, A.: Über Gravitationswellen. *Sitzungsberichte der Königlich Preußischen Akademie der Wissenschaften (Berlin), Seite* 154–167 (1918)
14. Misner, C.W., Thorne, K.S., Wheeler, J.A.: Gravitation. Freeman, W. H (1973)
15. Burke-Spolaor, S., et al.: The astrophysics of nanohertz gravitational waves. A&ARv **27**, 5 (2019)

16. The LIGO Scientific Collaboration et al.: Advanced LIGO. Class. Quant. Gravity **32**, 074001 (2015)
17. Fairhurst, S.: Improved source localization with LIGO-India. J. Phys: Conf. Ser. **484**, 012007 (2014)
18. Key, J.S., Han, R.J., Ara, S.R.: For the Cosmic Explorer project. Cosmic explorer observatory conceptual design. J. Phys.: Conf. Ser. **3177**, 012098 (2026)
19. Punturo, M., et al.: The Einstein telescope: a third-generation gravitational wave observatory. Class. Quant. Gravity **27**, 194002 (2010)
20. Abac, A., et al.: The science of the Einstein telescope. J. Cosmol. Astropart. Phys. **2026**, 081 (2026)
21. Jenet, F.A., Romano, J.D.: Understanding the gravitational-wave Hellings and Downs curve for pulsar timing arrays in terms of sound and electromagnetic waves. Am. J. Phys. **83**, 635–645 (2015)
22. Hellings, R.W., Downs, G.S.: Upper limits on the isotropic gravitational radiation background from pulsar timing analysis. ApJL **265**, L39–L42 (1983)
23. Tarafdar, P., et al.: The Indian pulsar timing array: first data release. Publ. Astron. Soc. Aust. **39**, e053 (2022)
24. EPTA Collaboration, et al.: The second data release from the European Pulsar Timing Array. I. The dataset and timing analysis. A&A **678**, A48 (2023)
25. EPTA Collaboration, et al.: The second data release from the European Pulsar Timing Array. III. Search for gravitational wave signals. A&A 678, A50 (2023)
26. Agazie, G., et al.: The NANOGrav 15 yr data set: evidence for a gravitational-wave background. ApJL **951**, L8 (2023)
27. Reardon, D.J., et al.: Search for an isotropic gravitational-wave background with the Parkes Pulsar Timing Array. ApJL **951**, L6 (2023)
28. Xu, H., et al.: Searching for the nano-hertz stochastic gravitational wave background with the Chinese Pulsar Timing Array data release I. Res. Astron. Astrophys. **23**, 075024 (2023)
29. Nobleson, K.: The Indian pulsar timing array data release 2: II. Customised single-pulsar noise analysis and noise budget. J. High Energy Astrophys. **53**, 100594 (2026)
30. Iraci, F., et al.: Combining the second data release of the European Pulsar Timing Array with low-frequency pulsar data. A&A **704**, A109 (2025)
31. Joshi, B.C., Karastergiou, A., Burgay, M.: Pulsar science with the SKA observatory. Open J. Astrophys. **8**, 54638 (2025)
32. Shannon, R.M., et al.: The SKAO Pulsar Timing Array. Open J. Astrophys. **8**, 54243 (2025)

The Dark Matter Direct Search Initiative at JUSL, India

Mala Das[1,2(✉)] and Kaynat Fatima[1,2]

[1] NAPP Division, Saha Institute of Nuclear Physics, 1/AF Bidhannagar, Kolkata 700064, India
mala.das@saha.ac.in

[2] Homi Bhabha National Institute, Training School Complex, Anushakti Nagar, Mumbai 400094, India

Abstract. The Indian Dark matter search Experiment (InDEx) aims to search for low-mass Weakly Interacting Massive Particles (WIMPs) using superheated liquid detectors (SLD) operated at the Jaduguda Underground Science Laboratory (JUSL). SLD with active liquid $C_2H_2F_4$ was used in the experiment. Run-1 and run-2 of InDEx have been completed at thresholds of 5.87 keV and 1.95 keV, respectively. The corresponding minimum WIMP mass sensitivity of $C_2H_2F_4$ is 4.5 GeV/c^2 for run-1 and 2.3 GeV/c^2 for run-2. Run-2, with 7.2 kg.days exposure sets constraint on spin-independent cross section at $(1.55 + (+0.62–0.32)_{stat} + (+0.03–0.02)_{sys}) \times 10^{-40}$ cm^2(90% C.L.) and on spin dependent cross section on F at $(7.97 + (+3.44–1.78)_{stat} + (+0.15–0.18)_{sys}) \times 10^{-38}$ cm^2 (90% C.L.) for WIMP masses of 20.4 GeV/c^2 and 21.0 GeV/c^2, respectively.

Keywords: dark matter · weakly interacting massive particles · superheated liquid detectors

1 Introduction

The existence of dark matter (DM) is strongly supported by several astrophysical and cosmological observations, however, its particle nature remains unknown [1, 2]. Weakly Interacting Massive Particles (WIMP) are one of the favoured candidates of predicted DM [3]. Direct detection experiments attempt to observe nuclear recoils resulting from elastic scattering of DM particles with detector nuclei. While many experiments have focused on the GeV-TeV WIMP mass range, the null results have motivated increased interest in the low-mass DM regime. In the low-mass region, the expected nuclear recoil energies are small, making their detection experimentally challenging. This has led to the development of detector technologies with intrinsically low energy thresholds and strong background rejection capabilities. In this context, the Indian Dark matter search Experiment (InDEx) has been initiated at the Jaduguda Underground Science Laboratory (JUSL), Jharkhand, India. The 555 m rock overburden (~1604 m water equivalent) of JUSL provides shielding against cosmic-ray induced backgrounds [4]. InDEx employs

Mala Das; Kaynat Fatima et al for InDEx collaboration

G. B. Mohanty and V. S. Prabhudesai (Eds.): DAEC 2026, SPPHY 368, pp. 168–172, 2026.
https://doi.org/10.1007/978-981-92-2948-2_21

superheated liquid detectors (SLD) containing active liquid $C_2H_2F_4$ (b.p. -26.3 °C) [5–8]. SLD offers a powerful approach for low-mass DM searches due to their intrinsic threshold behaviour, which allows them to be insensitive to minimum ionizing radiation while remaining sensitive to higher linear energy transfer nuclear recoils. By tuning the operating temperature and pressure, the recoil energy threshold can be precisely controlled [9]. The presence of hydrogen, possible lowest mass target nucleus in $C_2H_2F_4$ allows InDEx to extend its sensitivity toward the low-mass WIMP regime [9]. So far, run-1 and run-2 of the InDEx experiment have been completed at thresholds of 5.87 keV and 1.95 keV, respectively, with total exposures of 2.47 and 7.2 kg·days. The estimated minimum WIMP mass sensitivity of $C_2H_2F_4$ is 4.5 GeV/c^2 for run-1 and 2.3 GeV/c^2 for run-2 [7, 8]. In this article, we report the experimental set up and analysis results from run-1 and run-2 of the InDEx experiment. We also outline future plans aimed at improving sensitivity, understanding the gamma-ray induced backgrounds, and scaling up the experiment.

2 Experimental Details

The $C_2H_2F_4$ SLDs were fabricated at the laboratory of Saha Institute of Nuclear Physics using a pressure reactor. The SLD consists of the droplets of superheated liquid suspended in aquasonic gel matrix. Fabricated emulsion was housed inside a 500 ml borosilicate glass container. The SLD was connected with a piezoelectric acoustic sensor on top of it touching the aquasonic gel. The output of the acoustic sensor was connected to FPGA-based data acquisition system and the signals were recorded using LabVIEW software [10]. During data acquisition, the amplitude threshold was set above the electronic noise level.

For run-1 of the InDEx a single SLD was installed at JUSL and operated at room temperature (25.4 °C), corresponding to a threshold of 5.87 keV. The effective data taking period of run-1 was 48.6 days. The detector contained (50.81 $\pm$ 0.01) gm of active liquid, resulting in a total 2.47 kg.days of exposure [7].

The run-2 of InDEx was carried out at a lower threshold of 1.95 keV, achieved by operating the detectors at a temperature of 35° C. Two SLDs, containing total of (70.40 $\pm$ 0.01) gm active liquid, were installed inside a temperature controlled SS system (TCS). Each detector was instrumented with an independent piezoelectric sensor, and the acoustic signals were recorded separately in two data acquisition channels. The detectors and experimental setup of run-2 at JUSL is shown in Fig. 1. The effective live time of run-2 was 102.48 days, resulting in a total exposure of 7.2 kg.days [8].

The background studies at JUSL indicate that neutrons and gamma-rays constitute the dominant radiation backgrounds at the experimental site [4]. Previous measurements have shown that $C_2H_2F_4$ SLD become sensitiveto gamma-rays above operating temperature of (38 $\pm$ 1.4) °C [11, 12]. Consequently, both run-1 and run-2 detectors are insensitive to background gamma-rays and hence neutrons become the dominant background for the experiment. Calibration measurements were performed at SINP using a ^{241}AmBe neutron source (10 mCi) at both operating thresholds of 5.87 and 1.95 keV to characterize the detector response to nuclear recoils.

(a) (b)

Fig. 1 (a) Detectors and (b) experimental setup of InDEx run-2 at JUSL.

3 Results and Discussion

Data recorded during run-1 and run-2 of the InDEx experiment were analyzed to identify the true bubble nucleation events and to reject noise-induced triggers. Several parameters such as number of peaks (N_{peaks}), signal rise time (R_t), peak amplitude (A_P) etc. were constructed for each of the recorded events of InDEx runs and also for calibration runs. The distribution of the selected parameters for the InDEx data were found to be consistent with those obtained from calibration measurements [7, 8].

The expected neutron background rates were estimated for run-1 and run-2 using the rate equation as described in ref. [5], accounting for contributions from both radiogenic and cosmogenic neutrons at JUSL. The spectrum and flux of neutron for rate calculation have been considered from Ref. [4]. The estimated neutron-induced event rate for run-1 is $(2.22 \pm 0.20 \text{ stat}) \times 10^{-6}\ s^{-1}\ g^{-1}$, while for run-2 is $(1.91 \pm 0.10 \text{ stat}) \times 10^{-7}\ s^{-1}\ g^{-1}$ [7]. The experimentally observed event rates are $(2.06 \pm 0.10 \text{ stat}) \times 10^{-6}\ s^{-1}\ g^{-1}$ for run-1 and $(1.36 \pm 0.14 \text{ stat}) \times 10^{-7}\ s^{-1}\ g^{-1}$ for run-2 [8].

Since no excess of events over the expected background was observed in either run, upper limits on the WIMP-nucleon interaction cross-sections were derived with 90% confidence level (C.L.) [9, 13, 14]. The exclusion limits (at 90% C.L.) in the parameter space of spin-independent (SI) WIMP-nucleon cross-sections for run-1 and run-2 are shown in Fig. 2.

Run-1, with 2.47 kg.days of exposure, sets a constraint on SI cross-section of (8.47 + $(+0.38\text{–}0.42)_{stat}$ + $(+0.04\text{–}0.05)_{sys}$) $\times 10^{-39}$ cm^2 (90% C.L.) for a WIMP mass of 31.5 GeV/ c^2. A corresponding constraint on the spin-dependent (SD) cross-section is obtained as ($3.25 \pm 0.15_{stat}$ + $(+0.01\text{–}0.02)_{sys}$) $\times 10^{-36}$ cm^2 (90% C.L.) for a WIMP mass of 31.8 GeV/c^2 [7].

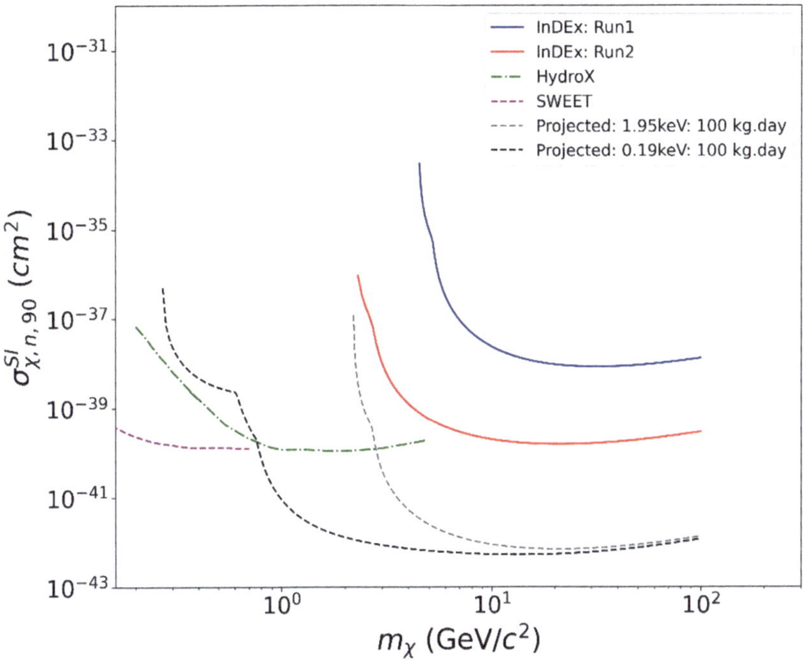

Fig. 2 Present and projected SI sensitivity of InDEx

Run-2, with an exposure of 7.2 kg.days, sets a more stringent constraint on SI cross-section at $(1.55 + (+0.62–0.32)_{stat} + (+0.03–0.02)_{sys}) \times 10^{-40}$ cm^2 (90% C.L.) for a WIMP mass of 20.4 GeV/c^2. The SD constraint on F is obtained at $(7.97 + (+3.44–1.78)_{stat} + (+0.15–0.18)_{sys}) \times 10^{-38}$ cm^2 (90% C.L.) for a WIMP mass of 21.0 GeV/c^2 [8].

The projected sensitivities of InDEx for zero background consideration are also shown in Fig. 2 at thresholds of 1.95 and 0.19 keV for a total exposure of 100 kg.days along with the projected limits of the hydrogen based experiments, HydroX and SWEET [15, 16].

4 Future Directions

Future phases of the InDEx experiment are planned with increased detector exposure in order to improve sensitivity to low-mass dark matter interactions. Scaling up the active mass and extending the data-taking period will enhance the statistical reach of the experiment. In addition to increased exposure, active shielding systems are being planned to further suppress ambient backgrounds. Such shielding, combined with optimized detector operation, is expected to reduce the background contributions significantly. Detailed analysis of InDEx data acquired in the gamma-sensitive threshold regime is currently in progress to better understand the detector response to background gamma-rays. These studies are essential for validating background models at low energy thresholds. Furthermore, systematic investigations of the intrinsic radioactive background of detector materials are underway. Improved characterization of intrinsic backgrounds will play a crucial role in refining the background estimates and guiding future detector design. With increased exposure, active background suppression, and a deeper understanding of both external and intrinsic background sources, future runs of the InDEx experiment are expected to achieve improved sensitivity in the low-mass dark matter parameter space.

References

1. D'Amico, G., et al.: arXiv: 0907.1912. (2009)
2. Roszkowski, L., et al.: Rep. Prog. Phys. **81**, 066201 (2018)
3. Goodman, M.W., Witten, E.: Phys. Rev. D. **31**, 3059–3063 (1985)
4. Ghosh, S., et al.: Astropart. Phys. **139**, 102700 (2022)
5. Sahoo, S., et al.: Nucl. Instrum. Methods Phys. Res., Sect. A. **1008**, 165450 (2021)
6. Kumar, V., et al.: arXiv:2501.02833. (2025)
7. Das, S., et al.: Eur. Phys. J. Spec. Top. (2025). https://doi.org/10.1140/epjs/s11734-025-01889-1
8. Das, S., et al.: Phys. Rev. D. **112**, 042003 (2025)
9. Seth, S., et al.: Phys. Rev. D. **101**, 103005 (2020)
10. Sahoo, S., et al.: Nucl. Instrum. Methods Phys. Res., Sect. A. **1009**, 165457 (2021)
11. Sahoo, S., et al.: Nucl. Instrum. Methods Phys. Res., Sect. A. **931**, 44 (2019)
12. Fatima, K., et al.: Springer Proceedings in Physics, vol. 432. Springer, Singapore (2025). https://doi.org/10.1007/978-981-95-1513-4_19
13. Chaudhury, S., et al.: J. Cosmol. Astropart. Phys. **2010**(09), 020 (2010)
14. Cowan, K., et al.: Eur. Phys. J. C. **73**, 2501 (2013)
15. Lippincott, W.H., et al.: Coms. Phys. **8**, 244 (2025)
16. Bento, A., et al.: arXiv: 2510.00068v1. (2025)

Highest Energy Observations in Multi-messenger Astronomy

K. K. Singh[1,2(✉)]

[1] Astrophysical Sciences Division, Bhabha Atomic Research Centre, Mumbai 400085, India
[2] Homi Bhabha National Institute, Mumbai 400094, India
kksastro@barc.gov.in

Abstract. Multi messenger astronomy has now become a major avenue to explore the universe after decades of successive developments in astrophysical observations and phenomenology. It uses complementary pieces of information provided by different messengers, like photons, cosmic rays, gravitational waves, and neutrinos emitted from well-known sources, to derive a complete phenomenological picture of an astronomical event. These messengers also help in measuring the key cosmological parameters of the universe. In this contribution, an overview of Indian efforts in high-energy gamma-ray and cosmic-ray measurements within the emerging discipline of multi-messenger research is discussed. Given the recent developments in the field, this contribution focuses in particular on the ground-based gamma-ray astronomy with MACE at Hanle and cosmic-ray particle detection with GRAPES-3 at Ooty.

Keywords: Multi messenger astronomy · High Energy · Ground-based observations

1 Introduction

Multi-messenger astronomy (MMA) is emerging as a distinct and powerful tool to probe the extreme processes of the physical universe. Cosmic rays, photons, neutrinos and gravitational waves, referred to as the messengers of the universe, provide complementary information about different astrophysical processes [1]. Going beyond light in MMA allows us to address some of the fundamental questions in astro-particle physics such as relativistic particle acceleration, synthesis of heavy elements in stellar atmosphere, existence of matter and anti-matter, aspects of the cosmology, and physics of the violent non-thermal processes in the universe.

The all particle cosmic ray spectrum measured by various ground-based experiments world-wide spans over 13 orders of magnitude in energy ($\sim 10^9$eV – 10^{21}eV) and 30 orders of magnitude in flux [2]. The featureless differential energy spectrum of cosmic rays, approximately described by a power law, indicates the presence of powerful particle accelerators in Galactic and extragalactic space.

G. B. Mohanty and V. S. Prabhudesai (Eds.): DAEC 2026, SPPHY 368, pp. 173–176, 2026.
https://doi.org/10.1007/978-981-92-2948-2_22

Modern astronomical instruments or telescopes are capable of detecting the photons of electromagnetic radiation in a very wide energy range from radio to very high energy gamma rays emitted by astrophysical sources [3,4]. The highest energy end of electromagentic spectrum, characterizing the cosmic gamma radiation, spans from MeV to PeV energies. Following the first detection of solar neutrinos by the underground neutrino detectors in 1968, about 20 neutrinos of MeV energies were detected from the core-collapse supernova 1987A. The current decade marked the first breakthrough with the discovery of extragalactic neutrinos in the TeV–PeV energy range [5,6]. The first discovery of gravitational waves using Laser Interferometers in 2015 [7] presented a completely new means for observational astronomy and confirmed the true presence of multi-messengers in the universe.

The first observation of gravitational waves from a binary neutron star merger event GW170817 in 2017 and its subsequent association with the gamma ray burst GRB 170817A, detected by the space-based Gamma Ray Burst Monitor onboard *Fermi* satellite 1.7 s after the coalescence, marked the beginning of the golden era of MMA and time-domain astronomy [8]. GW170817 remains the only multi-messenger event associated with the gravitational and electromagnetic radiation together so far. However, the ensemble of current and upcoming state-of-the-art observational facilities around the globe are expected to generate an unprecedented volume of triggers for MMA [9]. This contribution mainly highlights the recent gamma ray and cosmic ray measurements from the Indian ground-based observatories: MACE (Major Atmospheric Cherenkov Experiment) and GRAPES-3 (Gamma Ray Astronomy at PeV EnergieS phase-3) and their future prospects.

2 The MACE Telescope

MACE is a key observational facility to study the high energy gamma rays of cosmic origin. It is situated at an altitude of $\sim$ 4.3 km in Hanle, Ladakh India [10,11]. The geographical location of MACE, shown in Fig. 1(left), in the East provides an important longitudinal advantage with respect to the major gamma ray observatories in the West. Equipped with a 21 m diameter quasi-parabolic light corrector, MACE belongs to the class of large sized imaging atmospheric Cherenkov telescopes such as the MAGIC, LST, and H.E.S.S. II, operating in different corners of the world. The design and site of MACE closely follow the idea of having a lower energy threshold gamma ray telescope at high altitude to explore the universe in the high energy energy above 20 GeV. The telescope had first light in April 2021 followed by its successful commissioning in September 2021 for regular science observations [12].

MACE has achieved important scientific milestones in just four years of its science operation [13]. The energy spectrum of high energy photons, detected from the standard candle Crab Nebula, has been derived in the energy band 80 GeV–5 TeV. The telescope has detected giant gamma-ray flares from the radio galaxy NGC 1275 at three different epochs between December 2022 and January

2025. These measurements have been used to study the broadband spectral energy distribution of the source thereby constraining the jet viewing angle below 30°. The MACE observations have also been used to place maiden constraints on the integral flux above 70 GeV for a number of high redshift blazars. The telescope has successfully detected gamma-ray flaring episodes from well-studied sources like Mrk 421, Mrk 501, 1ES1959+650 and OP 313. Detection of signal from the most distant blazar OP 313 with MACE has a special significance as it pushes the cosmic gamma ray horizon close to a redshift ~ 1 underlining the low threshold energy potential of the telescope. It is proposed to install two more MACE like telescopes at Hanle to establish a Stereoscopic MACE System (SMS). This will create a unique stereoscopic gamma ray observatory at probably the highest altitude. In the next phase, SMS will be complemented by an array of 4 m class small-sized imaging Cherenkov telescopes to increase the energy coverage up to 100 TeV by 2035.

Fig. 1. A view of the MACE telescope situated at Hanle, UT of Ladakh (left) and the GRAPES-3 experiment at Ooty, Tamil Nadu (right) in India.

3 GRAPES-3

The GRAPES-3 experiment is located at an altitude of 2.2 km in Ooty, Tamil Nadu, India. It has been operationl since 2001. As depicted in Fig. 1 (right), the current GRAPES-3 experiment consists of an array of 400 plastic scintillators, each with an area of $\sim 1\,\text{m}^2$ at 8 m separation, covering a total area of $\sim 25{,}000\,\text{m}^2$ on the ground and an underground proportional counter based muon tracker of area $\sim 560\,\text{m}^2$ [14,15]. This facility has been set up to probe the origin, acceleration and propagation of cosmic rays in the energy range from 10 TeV to 10 PeV. It is also being used to study various phenomena associated with the solar activity and thunderstorm through cosmic muons. The recent most important science highlights of GRAPES-3 are discovery of a kink in the cosmic ray proton spectrum at around 166 TeV, identification of small-scale anisotropic structures in the arrival distribution of cosmic rays, measurement of the electrical properties of a thundercloud using muon imaging, and probing the transient weakening of Earth's magnetic shield. In the coming times, the scintillation array will be upgraded to a coverage area of $\sim 80{,}000\,\mathrm{m}^2$ and muon telescope tracker area will be extended $\sim 1130\,\mathrm{m}^2$. A new facility for radio measurements from cosmic ray showers will also be developed.

4 Summary

The substantial progress made over the last decade has established MMA as an observational discipline in its own right. A combination of information provided by the detection of all four fundamental forces of nature, namely the strong, weak, electromagnetic, and gravity, provides a formidable tool to probe the mysteries of the universe. Therefore, the current experimental facilities in India for high energy observations such as MACE and GRAPES-3 and their future augmentation have excellent prospects for making significant contributions in the emerging field of multi-messenger astronomy.

References

1. Mészáros, P., Fox, D.B., Hanna, C., Murase, K.: Multi-messenger astrophysics. Nat. Rev. Phys. **1**(10), 585–599 (2019)
2. Aab, A., Pierre Auger Collaboration et al.: Measurement of the cosmic-ray energy spectrum above 2.5×10^{18} eV using the Pierre Auger Observatory. Phys. Rev. D **102**, 062005 (2020)
3. Aharonian, F., Buckley, J., Kifune, T., Sinnis, G.: High energy astrophysics with ground-based gamma ray detectors. Rep. Prog. Phys. **71**, 096901 (2008)
4. Middleton, M.J., et al.: Paving the way to simultaneous multi-wavelength astronomy. New Astron. Rev. **79**, 26–48 (2017)
5. Aartsen, M.G., et al.: First observation of PeV-energy neutrinos with IceCube. Phys. Rev. Lett. **111**, 021103 (2013)
6. Aartsen, M.G., et al.: Multimessenger observations of a flaring blazar coincident with high-energy neutrino IceCube-170922A. Science **361**, 6398 (2018)
7. Abbott, B.P., et al.: Observation of gravitational waves from a binary black hole merger. Phys. Rev. Lett. **116**, 061102 (2016)
8. Abbott, B.P., et al.: Multi-messenger observations of a binary neutron star merger. ApJL **848**, L12 (2017)
9. Nissanke, S.: Multi-messenger and Time-Domain Astronomy in the 2040s. arXiv:2512.14546 (2025)
10. Singh, K.K., Yadav, K.K.: 20 years of Indian gamma ray astronomy using imaging Cherenkov telescopes and road ahead. Universe **7**, 96 (2021)
11. Singh, K.K.: Gamma-ray astronomy with the imaging atmospheric Cherenkov telescopes in India. J. Astrophy. Astron. **43**, 3 (2022)
12. Yadav, K.K., et al.: Commissioning of the MACE gamma-ray telescope at Hanle, Ladakh. India. Curr. Sci. **123**, 1428 (2022)
13. Singh, K.K.: MACE: a unique national facility for astrophysics research at very high energy and future prospects. Proc. Ind. Nat. Sci. Acad. **92**, 28–33 (2026)
14. Gupta, S.K., et al.: GRAPES-3—a high-density air shower array for studies on the structure in the cosmic-ray energy spectrum near the knee. NIMPA A **540**, 311–323 (2005)
15. Hayashi, Y., et al.: A large area muon tracking detector for ultra-high energy cosmic ray astrophysics—the GRAPES-3 experiment. NIMP A **545**, 643–657 (2005)

Precision, Discovery, and the Quest for New Physics at Colliders

Aruna Kumar Nayak(✉)

Institute of Physics, Bhubaneswar, India
nayak@iopb.res.in

Abstract. A Higgs boson was discovered by the ATLAS and CMS experiments at CERN-LHC, ensuring the validity of the electroweak symmetry breaking mechanism in generating the masses of the fundamental fermions and gauge bosons. However, LHC has not found yet any new particles predicted by several beyond Standard Model theories, which have been proposed to address several inconsistencies in the Standard Model. In this scenario, the particle physics communities have been studying various road maps for the progress of the particle physics in future to understand the fundamental questions of nature. This article briefly presents the current status and future directions while discussing the efforts of the Indian collider based experimental physics community, especially from the institutes under Department of Atomic Energy.

Keywords: LHC · HL-LHC · ALICE · CMS · BELLE-II · Higgs · BSM

Particle physics is at crossroads following the discovery of a Higgs boson by the ATLAS and CMS experiments at CERN-LHC [1,2], which completed the Standard Model (SM) of particle physics. Prior to this discovery all other particles predicted by the SM were discovered, the last being the top quark by the experiments at Tevatron, Fermilab. The particle colliders have played a leading role in these pursuits. Following the discovery of the Higgs boson determining its properties has been a major focus of the LHC physics program [3,4]. These include measuring the couplings to fermions and gauge bosons, spin and parity, width of the Higgs boson, and its self coupling. Most of the properties measured so far are compatible with that expected from the SM. Figure 1 shows the measured coupling modifiers of the Higgs boson to fermions and gauge bosons, as measured by the CMS experiment using the resolved coupling modifier fit [5]. The result shows agreement with the SM predictions over several orders of magnitude in mass. Furthermore, the experiments have been carrying out measurements for many SM processes and parameters, including that of top quarks. No major deviations from SM predictions have been observed.

Despite enormous success of the SM in describing the elementary particles and their interactions at the present energy scale it is believed to be incomplete. There are several experimental observations that can not be described

G. B. Mohanty and V. S. Prabhudesai (Eds.): DAEC 2026, SPPHY 368, pp. 177–181, 2026.
https://doi.org/10.1007/978-981-92-2948-2_23

by the SM. For example, it does not explain the nature of dark matter, which constitutes about 26.8% of total mass-energy of the Universe. It does not provide a suitable explanation for the matter-antimatter asymmetry observed in the nature. Similarly, the recent observations of neutrino flavor oscillations, suggesting neutrinos have a tiny mass, can not be accommodated within the framework of SM, which treats neutrinos to be massless. Furthermore, there are several theoretical inconsistencies, such as mass hierarchy problem [6], (non)unifications of gauge couplings at higher energy scale. Several extensions of the SM have been proposed, which seek to address many of the above inadequacies of the SM [7]. Most of these models predict existence of new particles, including that of additional Higgs bosons. The experiments at LHC are carrying out direct searches for beyond SM (BSM) particles. There is no concrete evidence yet for the presence of any BSM particles within the reach of the LHC energy. The results of the searches using data from LHC experiments are ruling out large region of the parameter spaces in these proposed models. However, the LHC experiments have so far analyzed about 150 fb^{-1} of pp collision data, which were collected during Run-2 operations of LHC from 2016 to 2018. LHC is delivering about 350 fb^{-1} of data during its Run-3 operations from 2022 to 2026. Most of these data are yet to be fully analyzed.

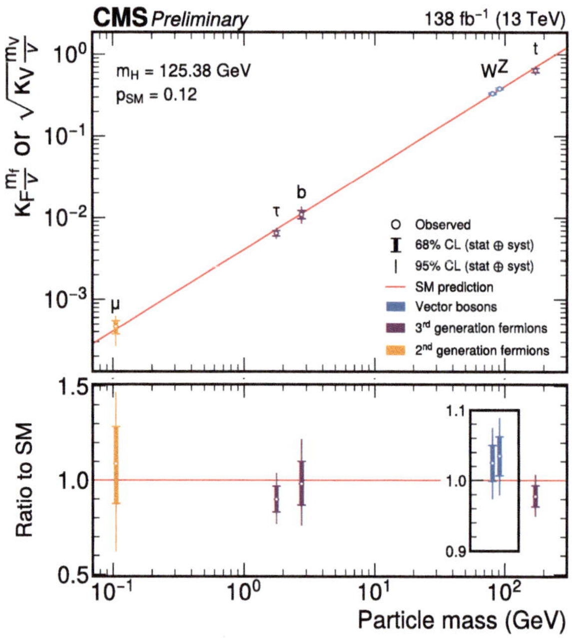

Fig. 1. The measured coupling modifiers of the Higgs boson to fermions and gauge bosons by the CMS experiment [5].

As there is no signature of any new physics yet, the particle physics community is adopting two complementary approaches: (1) Direct searches for new particles as predicted by the models and (2) Precision measurements of key SM parameters, especially the Higgs, top quark, electroweak, and flavor sectors, which are sensitive to the presence of BSM particles at higher energy scales. For example, the effect of new particles from BSM theory can make a small corrections to the Yukawa couplings of the fermions, which would make a small deviation in their values with respect to what is expected from SM. Although, the values of measured couplings at LHC are in agreement with the predictions of the SM within their uncertainties, these are not yet precise enough to observe such a deviation expected from contributions of BSM particles at higher energy scales. It is essential to measure the Yukawa couplings to a very high precision ($\mathcal{O}(1\%)$) to observe any possible deviation. The other important parameter is the trilinear self coupling parameter (λ_3) of the Higgs boson, which provides key information about the shape of the Higgs potential. This can be measured through the production of a pair of Higgs boson. However, the production cross section of this process is about three orders of magnitude smaller than that of the production of single Higgs boson. Therefore, it requires a large amount of data to probe this parameter. The current result for the measurement of λ_3 parameter from ATLAS and CMS experiments is shown in Fig. 2(left). The LHC is going to be upgraded to high luminosity LHC (HL-LHC) to deliver about 3000 fb^{-1} pp collision data. The expected results from HL-LHC may bring the uncertainties on some of the coupling measurements to 2–4%. The HL-LHC projection of the measurement of λ_3 parameter is shown in Fig. 2(right). Combining the results from the two experiments, the uncertainties on λ_3 might go down to below 30%.

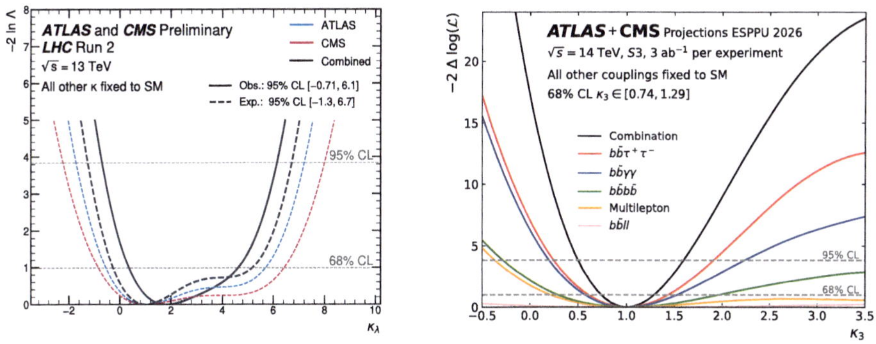

Fig. 2. Left: The combined measurement of self coupling modifier by ATLAS and CMS experiments using pp collisions at 13 TeV, taken from Ref. [8]. Right: The expected measurement of self coupling modifier projected to 3000 fb^{-1} of data, taken from Ref. [9].

In this scenario, the international particle physics community is discussing various strategies and coming up with road maps for future directions [10,11].

Several low energy e^+e^- precision collider machines, such as CLIC, ILC, FCC-ee, and CEPC, are being considered along with options for multi-TeV $\mu^+\mu^-$ collider and very high energy ($\approx$ 85 TeV) hadron-hadron machine (FCC-hh). The precision machines are expected to constrain the Higgs boson couplings to particles to a percent level or smaller. The constraint on λ_3 parameter can reach about 10%. With FCC-hh it can go down further to a few percent level.

India has a rich history of research into the understanding of fundamental laws of the universe. It has always been in the forefront of the research in particle physics, both in theoretical as well as in experimental aspects. In the past Indian scientists have played pioneering role in performing cosmic ray experiments on the surface of earth, deep underground, as well as in the upper atmosphere by flying detectors using balloons. Several institutes have been involved in collider experiments at leading world laboratories, such as CERN, Fermilab, KEK, and BNL, contributing to the discoveries of top quark and Higgs boson, CP violation in B-meson decays, as well as to the precision measurements of the electroweak and QCD parameters. The collider based experimental community has grown significantly in the past few decades. Currently, Indian institutes are collaborating with ALICE and CMS experiments at CERN-LHC, STAR and PHENIX experiment at RHIC, BNL, and BELLE-II experiment at KEK, JAPAN. These include institutes under Department of Atomic Energy (DAE): Bhabha Atomic Research Center (BARC) and Tata Institute of Fundamental Research (TIFR) in Mumbai, Saha Institute of Nuclear Physics (SINP) and Variable Energy Cyclotron Center (VECC) in Kolkata, National Institute for Science Education and Research (NISER) and Institute of Physics (IoP) in Bhubaneswar. BARC and TIFR had made significant contributions to the current CMS detector. TIFR was the primary contributor to the outer hadron calorimeter (HO) of the CMS detector, with the assembly, testing, and installation of the plastic scintillator tiles. About 1100 silicon sensors for the ECAL preshower detector was fabricated in India with the fruitful collaboration of BARC and Bharat Electronics Ltd (BEL). Several RPC and GEM chambers for CMS muon detectors were assembled, tested, and delivered by BARC. Furthermore, TIFR has made major contributions to the assembly of silicon vertex detectors in BELLE-II experiment [12]. The phase-II upgrade of the CMS detector for HL-LHC has provided immense opportunities for Indian groups to contribute to the technologically challenging components of the detector. DAE institutes are contributing to the building of outer silicon-strip tracker [13], silicon sensor based High granularity calorimeter (HGCAL) [14], and Level-1 trigger system [15]. In this effort NISER, IoP, and SINP are working together for assembly, testing, and integration of about 1100 silicon-strip 2S modules. A module assembly center has been established at NISER. Several prototype and pre-production modules have been already assembled and successfully tested. The center is working in collaboration with other tracker module assembly centers of CMS to deliver required number of modules for installation into the CMS detector. TIFR, along with SINP, is playing a significant role in building of the HGCAL detector. TIFR and SINP carried out extensive R&Ds for the fabrication of low density hexaboard PCBs in India for HGCAL front-end electronics, which enabled industries to fabricate it in India for supplying to CERN. TIFR, in collaboration with Indian industries, is delivering about 113 Serinity-S boards, which are high performance ATCA platforms designed to handle the vast data volumes and high demanding trigger requirements. These boards are required for DAQ path of the HGCAL. Additional electronics components, such as multiphase DC-DC boards for backend electronics, have also been developed by TIFR. Furthermore, TIFR is establishing a module assembly center to assemble a large fractions of the HGCAL modules, perform functional tests, and deliver them for installation to the CMS detector.

In summary, the particle physics experiments in energy frontier are entering an era of precision physics. The Indian particle physics experimental communities have long been involved in several international collider based experiments, contributing to detector building and physics analyses. The upcoming detector upgrades for HL-LHC has provided opportunities to participate in building of several technologically advanced detector components. The expertise gained through these involvements will benefit immensely to carry out R&Ds for future experiments.

References

1. ATLAS Collaboration: Phys. Lett. B **716**, 1 (2012). arXiv:1207.7214
2. CMS Collaboration: Phys. Lett. B **716**, 30 (2012). arXiv:1207.7235
3. ATLAS Collaboration: Nature **607**(7917), 52 (2022). Nature 612(7941), E24 (2022) (erratum). arXiv:2207.00092
4. CMS Collaboration: Nature **607**(7917), 60 (2022). arXiv:2207.00043
5. CMS Collaboration: CMS-PAS-HIG-21-018
6. Craig, N.: Eur. Phys. J. C **83**, 825 (2023). arXiv:2205.05708 [hep-ph]
7. Fox, P.J., et al.: arXiv:2210.03075 [hep-ph]
8. CMS Collaboration: CMS-PAS-HIG-25-014
9. CMS Collaboration: CMS-HIG-25-002. arXiv:2504.00672 [hep-ex]
10. Narain, M. et al.: arXiv:2211.11084 [hep-ex]
11. de Blas, J., et al.: CERN-ESU-2025-001. arXiv:2511.03883 [hep-ex]
12. Adamczyk, K., et al.: JINST **17**, P11042 (2022). arXiv:2201.09824
13. CMS Collaboration: CMS-TDR-014
14. CMS Collaboration: CMS-TDR-019
15. CMS Collaboration: CMS-TDR-021

Cutting-Edge Technology

Research with Ultra-High Intensity Laser Driven Plasma-Based Particle Accelerators and Betatron X-ray Source

Anand Moorti(✉)

Raja Ramanna Centre for Advanced Technology, Indore 452013, India
moorti@rrcat.gov.in

Abstract. Advancement in associated technologies has led to the availability of laser systems providing ultra-short (femto-second) pulse duration with Petawatt (PW) peak power and hence extreme focussed intensities of $> 10^{18-21}$ W/cm^2. Laser matter/plasma interaction at such extreme intensities involves fascinating physics and also applications, particularly towards research and development of Laser Plasma Accelerator (LPA). Due to associated large accelerating electric field (>GV/cm), three orders of magnitude higher than RF accelerator, LPA is seen as a potential technique to develop comparatively compact, and low-cost particle (electron/proton) accelerators. Extensive research and investigations performed at various facilities worldwide have shown acceleration of electrons upto ~ 10 GeV, and protons to 100 MeV energy using PW class lasers. Laser plasma electron accelerator driven x-ray/γ-ray sources and development of compact Free Electron Laser (FEL) are also being explored. In this talk, various aspects of laser driven plasma based advanced particle acceleration techniques, and international scenario in this area would be presented. National scenario, particularly, experimental investigations performed on LPA at RRCAT, Indore, showing acceleration of electrons to > 500 MeV energy, protons upto 20 MeV energy, and betatron x-ray source generation, using 150TW and 1 PW Ti:Sapphire lasers, along with future plan and possible applications would also be discussed.

Keywords: Petawatt class laser · laser plasma accelerator · electron acceleration · betatron x-ray source generation

Visualising the impact of particle (electron, proton / ion) accelerators on scientific research, technology development and enormous applications several high-energy accelerator facilities have been built worldwide. These particle accelerators uses radio-frequency (RF) based acceleration technique where the acceleration gradient is limited to $<$ 100 MV/m, due to the material breakdown on the walls of the RF accelerating cavities. Therefore, RF based high energy accelerators have large size and associated tremendous cost; consequently require multinational collaborations, and several years of construction time. Therefore, more advanced and efficient acceleration techniques of high acceleration gradient are required so that comparatively compact and low cost accelerators can be developed. This would lead to wider availability of accelerator facilities for variety of applications viz. Compact x-ray/γ-ray sources, medical applications e.g.

G. B. Mohanty and V. S. Prabhudesai (Eds.): DAEC 2026, SPPHY 368, pp. 185–189, 2026.
https://doi.org/10.1007/978-981-92-2948-2_24

in cancer therapy, sterilizing food, treating materials used in industry, and also disposing of nuclear waste etc.

To address the above limitation of RF accelerator, it was realized that 'plasma' can be used as a medium for particle acceleration, as it has no electrical breakdown limit, and therefore can support large accelerating field, which can be generated through charge separation. Ultra-short duration (femtosecond to picosecond range) laser pulse or charged particle beam can be sued to generate charge separation in a plasma medium.

Advancement in laser technology, has made availability of lasers suitable for particle acceleration, as laser pulse duration has been brought down to few femtoseconds (10^{-15} s), and laser pulse power in the pettawatt (PW) regime even with moderate laser pulse energy of tens of Joule per pulse. When such high-power laser pulses are focused to a tiny spot (few to tens of μms) very high intensities could be achieved i.e. > 10^{20} W/cm^2, causing plasma formation of variety of targets viz. Solid, liquid, gas, gaseous clusters etc. Inside a plasma the laser radiation can propagate only upto a certain electron density (known as critical density) where laser frequency (ω) equals the plasma frequency (ω_p). Based on the density regime laser plasma interaction can be classified into (1) overdense ($\omega < \omega_p$) e.g. solid target plasma where laser plasma interaction takes place in a small region close to critical density, and (2) underdense ($\omega < \omega_p$) e. g. low density gas targets where laser can propagate all through the plasma medium. Laser plasma interaction both in the overdense and underdense regime represents a physically rich topic on the interaction of electromagnetic radiation with matter, and hence involves several interesting physical processes and energy transport phenomena including relativistic and nonlinear effects. The several possible potential applications of laser driven plasma include: intense thermal and coherent x-ray/xuv ultra-short pulse generation, laser driven inertial confinement fusion and also laboratory astrophysics studies. In addition, advent of ultra-short (femtosecond), ultra-intense laser systems opened up an exciting area of laser plasma acceleration of electrons and protons.

A promising proposal towards laser plasma electron acceleration was given way back in 1979 by Tajima and Dawson [1], where it was shown that high-power, ultra-short laser pulse can drive high amplitude electron plasma wave in the wake of the laser pulse, and its intense Coulomb field can be used for electron acceleration. This scheme is known as "Laser Wake-field Electron Acceleration (LWFA)". In a typical laser wakefield accelerator, an intense laser beam is focused onto underdense plasma target of length few mms to several tens of cms (produced using gas-jets/gas-cells or preformed plasma channel in capillary discharge). The foot of the laser pulse ionizes the gas and forms plasma, and ponderomotive force associated with the main part of the laser pulse pushes electrons forward in the leading edge and backward in the trailing edge and hence leading to excitation of large amplitude relativistic plasma wave in the wake of the laser pulse. Self-injection of plasma electrons in the wave may occur due to wave breaking or ionisation induced injection processes which are trapped in the wave and accelerated to high energies (tens of MeV to few GeV based on laser and plasma length). Guiding of the laser pulse in plasma is achieved by relativistic self-focusing effect or due to parabolic density profile in preformed plasma channels which leads to propagation of laser pulse to several Rayleigh lengths. Maximum accelerating field gradient in a plasma is: $E = 0.96\ n_0^{1/2}$, where n_0 is the uniform plasma density. For $n_0 = 10^{18}$ cm^{-3},

E = 100 GV/m, i.e. thousands times larger than the conventional RF accelerators. This high acceleration gradient associated with the laser wakefield accelerators has paved the way for development of compact table-top electron accelerators, with unique features of ultrashort electron pulse duration (few fs) and kA peak currents, with an energy spread of about a few percent. In addition to various possible applications of electron beam, such an accelerator is also suitable for development of compact x-ray and γ-ray sources (by betatron radiation and inverse Compton scattering process), and x-ray free electron laser [2].

Subsequent to laser driven wakefield electron acceleration, it was also realized that a high-current, short-duration charged particle beam (electron/positron or proton) can also be used to generate electron plasma wave and hence to accelerate electrons [3]. This scheme is known as "Plasma Wake-field Acceleration (PWFA)". The Coloumb force due to charged particles space charge expels the plasma electrons (for electron beam)/pulls them (for positron/proton beam). The displaced electrons snap back to restore charge neutrality and overshoot their original position, setting up a plasma wave. A trailing particle beam behind the wakefield exciting beam is accelerated to high energies in short distances due to available high acceleration gradient wakefield. Major large accelerator laboratories e.g. SLAC, USA, and CERN, Geneva are working on PWFA scheme. SLAC has earlier demonstrated doubling of the energy of high-energy electron beam of 42 GeV to 85 GeV in a less than one meter of plasma. At CERN proton driven scheme is being investigated.

For proton/ion acceleration, the high-power laser is focused on a high-density, thin (sub-μm to few μm thickness) metallic foil plasma [4]. Interaction of ultra-short high intensity laser pulse with high-density solid plasma leads to generation of hot (fast) electrons of energy 100 s of keV to several tens of MeV due to various interaction and absorption processes in the laser propagation direction. The fast electrons escape rapidly from the rear surface of the thin metal targets, and proton/ion acceleration occurs at the rear surface of the metal foil due to huge sheath field (~ TV/m) set up by the escaping high-energy electrons. In this case ion acceleration is explained considering space charge field created by the hot electrons escaping from the rear surface of the target, and is known as Target Normal Sheath Acceleration (TNSA). Various other ion acceleration mechanism e.g. radiation pressure acceleration, collision less shock acceleration etc. are also being pursued in order to generate monoenergetic proton/ion beams. It may be mentioned here that, in case of laser plasma proton/ion acceleration filed gradient is of the order of MeVμm^{-1}, compared with $\sim$MeVm^{-1} in case of RF accelerators. The significant features of laser-driven proton beams are extremely small source size ($\sim$10 μm) of ion emission, and ultra-short duration (picoseonds at at the source). Drawbacks are the larger divergence (upto 10 s of degrees, and energy dependent) and the broad spectrum (generally ~ 100%). The beams are bright, with 10^{11}–10^{13} protons per pulse with energies $>$ MeV, corresponding to currents in the 100 s of kA range. However, the number of protons at the high-energy end of the spectrum can be lower.

At global level, impressive progress has been made in the area of laser driven plasma based acceleration during last two decades or so. On the electron acceleration front, suitable conditions for generation of high-energy ($>$ GeV), low-divergence (few mrad), quasi-monoenergetic (of energy spread 1–10%) electron beams have been identified.

Acceleration of electrons to 10 GeV in 30 cms long plasma, and protons to ~ 100 MeV using thin metal targets have been demonstrated using a PW Ti:sapphire lasers. Further, laser plasma electron accelerator based x-ray/γ-ray sources using betatron radiation and inverse Compton scattering process, and also compact Free-Electron Laser (FEL) have been developed and possible applications have been demonstrated. Still, several challenges remain. Reducing the energy spread to that demonstrated by RF accelerators is important, as is beam pointing and stability. Efforts are under way for generation of high-energy (50–250 MeV), proton beams with low divergence and energy spread, which has not been possible till date. Generation of high-quality (comparable to RF accelerators) beams is important for several applications. In this regard, suitable beam transport systems for LPA generated electron/proton beams are also being investigated and developed. Operation at high repetition rate would depend on the development of high-power ultra-short laser technology. Multi-stage laser plasma acceleration is also being investigated for developing high-energy accelerators, and may facilitate development of future particle colliders based on plasma acceleration technique.

In India, since the early days both experimental and theoretical studies on laser plasma interaction have been performed at various national labs and universities. BARC and TIFR, Mumbai have 200 TW laser system and at TIFR, Hyderbad centre a 1PW laser facility is also coming up. Extensive studies on laser plasma electron and proton acceleration have been performed at RRCAT, and high-power ultra-short Ti:Sapphire laser facilities have been set up which includes 150 TW, and recently installed 1 PW, 25 fs, 0.1 Hz, laser systems. Using gas-jet plasma of 4 mm length, acceleration of electron beams with quasi-monoenergetic spectrum with peak energy in the range of ~ 200 MeV to $\geq$ 500 MeV has been demonstrated, and role of Direct Laser Acceleration (DLA) along with Wakefield Acceleration (WFA) was shown [5]. Investigations on betatron x-ray generation with role of collective oscillations on flux enhancement and radiography and phase contrast imaging applications was also demonstrated [6]. Detailed investigations on JxB mechanism of acceleration of fast electrons in laser and thin metal foil interaction and on its associated key aspects were performed [7]. Experimental demonstration of acceleration of protons from few MeV to > 20 MeV energy, and its application for PET isotope production and nuclear reactions were also explored [8]. Experimental facilities is being set up using PW laser towards acceleration of electrons to $\geq$ 1 GeV and protons > 30 MeV, with suitable respective beam transport lines to explore possible applications of electron, proton beam and ultra-short x-rays.

Acknowledgments. Contributions of colleagues at Laser Plasma Division, RRCAT, in the investigations performed on laser plasma acceleration and x-ray source development are acknowledged.

References

1. Tajima, T., Dawson, J.M.: Laser electron accelerator. Phys. Rev. Lett. **43**, 267 (1979)
2. Albert, F.: Phys. Plasmas **30**, 050902 (2023)
3. Chen, P., et al.: Acceleration of electrons by the interaction of a bunched electron beam with a plasma. Phys. Rev. Lett. **54**, 693 (1985)
4. Macchi, A., et al.: Rev. Mod. Phys. **85**, 751–793 (2013)

5. D. Hazra et al., Plasma Phys. Control. Fusion **60**, 085015 (2018); ibid **61**, 125016 (2019); Phys. Rev. Accelerator Beams **22**, 074701 (2019); D. Hazra D, Ph. D. thesis, HBNI, DAE, Mumbai (2020). http://hbni.ac.in/thesis.html
6. S. Mishra et al., Phys. Rev. Acc. Beams **25**, 090703 (2022); Phys. Plasmas **31**, 103105 (2024)
7. T. Mandal et al., Phys. Plasmas, **30**, 023106 (2023); ibid **31**, 093102 (2024); T. Mandal, Ph.D. thesis, HBNI, DAE, Mumbai (2021). http://hbni.ac.in/thesis.html
8. M. Tayyab et al., Plasma Phys. Control. Fusion. **61**,115007 (2019); Nucl. Instrum. Methods Phys. Res. A **1065**, 169488 (2024); M. Tayyab, Ph.D. thesis, HBNI, DAE, Mumbai (2019). http://hbni.ac.in/thesis.html

An Infra-Red Free Electron Laser Based User Facility for Scientific Research & A Brief Overview of Accelerator Science and Technology Activities in the Department of Atomic Energy

Kamal Kumar Pant(✉)

Free Electron Laser and Utilization Section, Raja Ramanna Centre for Advanced Technology, Indore, M.P 452013, India
kkpant@rrcat.gov.in

Abstract. The **I**nfra-**R**ed **F**ree **E**lectron **L**aser (IR-FEL) at the Raja Ramanna Centre for Advanced Technology (RRCAT) delivers continuously tunable IR radiation in the 12.5–50 μm wavelength range with a maximum peak power > 3 MW and CW average power > 7 mW at ~ 20 μm wavelength at the experimental stations presently commissioned. This translates to a peak Brightness > 10^{17} photons/s/mm^2/mrad^2/0.1% BW and micro-pulse energy > 20 μJ in 10 ps pulses repeating at 29.75 MHz at the user stations over a wavelength range from 15–25 μm. The typical measured spectral width is ~ 1.4% at 22.7 μm wavelength. Good stability and repeatability of operation of the machine has been achieved. Six user ports have been built and two experimental stations are presently commissioned, serving users primarily from the condensed matter physics community. One more experimental station for IR irradiation studies is presently under development. Two ports for time domain studies will be developed in the near future to facilitate FEL pump - Optical probe and FEL pump - THz probe spectroscopy experiments. This article covers salient aspects of the features of the IR-FEL and its user facilities at RRCAT, Indore. A brief discussion on the possible directions for the future is also presented. An FEL is like a bridge between the science and technology of particle accelerators and lasers. This article also briefly presents a brief overview on the status of research and development activity related to particle accelerators in the Department of Atomic Energy.

Keywords: Tunable infra-red radiation · infra-red spectroscopy · short pulsed radiation · undulator · industrial accelerators · cyclotrons · superconducting accelerators · Radio Frequency Quadrupole

1 Introduction

A Free Electron Laser (FEL) is an electron accelerator based light source that delivers high peak power, coherent and monochromatic electromagnetic radiation with continuous wavelength tunability over its design wavelength range. By virtue of its principle of operation, an FEL can ideally be designed to lase at any required wavelength. Since an

G. B. Mohanty and V. S. Prabhudesai (Eds.): DAEC 2026, SPPHY 368, pp. 190–199, 2026.
https://doi.org/10.1007/978-981-92-2948-2_25

FEL generates lasing through a resonant interaction of a relativistic, short pulsed electron beam with a photon beam inside a spatially periodic arrangement of magnets in a device called as an undulator, the properties of FEL radiation depend critically upon the properties of the electron beam and the undulator. The wavelength of radiation generated by an FEL is given by the expression

$$\lambda_R = \frac{\lambda_U}{2\gamma^2}\left[1 + \left(\frac{eB_U\lambda_U}{2\pi mc}\right)^2\right]$$

where λ_R is the wavelength of the FEL radiation, λ_U is the undulator period, γ is the relativistic gamma factor, and $K_U = (eB_U\lambda_U)/(2\pi mc)$ is the undulator parameter, which depends upon the undulator magnetic field B_U and the undulator period λ_U. The terms 'm' and 'c' refer to electron rest mass and the speed of light in free space, respectively.

Tuning the wavelength of an FEL can be achieved by changing the energy of the electron beam (γ) or by changing K_U. While the former can be achieved by varying the microwave power fed to the linear accelerator (Linac) in the injector linac system of the FEL, it is not a fast process since a change in the electron beam energy necessitates a tuning of the electron beam transport line from the Linac exit up to the undulator to ultimately deliver the required electron beam quality at the undulator entry. For a fixed electron beam energy, however, fast tuning over a certain range of wavelength can be achieved by changing the undulator parameter K_U over its permissible range. For a fixed undulator period, K_U can be changed by varying the undulator gap, which results in change in B_U. FELs operating worldwide in the mid and far infra-red regions have demonstrated wavelength tunability by more than an order of magnitude by employing a combination of the above two methods.

Due to the significant interest worldwide in exploring new areas of science using intense radiation at long wavelengths in the mid/far IR and Tera Hertz (THz) regions of the electromagnetic spectrum, several FEL based user facilities have been developed with the first examples being the CLIO FEL in France, the FELIX FEL in Netherlands and the FEL at the Institute for FEL in Osaka (iFEL), Japan. Following these, many more IR/THz FELs and associated user facilities have been built in Russia, Japan, Germany, Netherlands, Korea, China and UK. Such FELs require modest electron beam energies from around 10 MeV to 50 MeV, and they have been built to deliver MW peak powers in picosecond pulses over a wavelength range from ~ 5 μm (60 THz) to 300 μm (1 THz). While the initial IR/THz FELs worldwide employed Normal Conducting (NC) Linac systems, some recent FELs employ Superconducting (SC) Linac systems. The move from NC Linacs to SC Linacs has resulted in a significant increase in the duty cycle of operation of these machines, and consequently in their operation with high Continuous Wave (CW) average power. The FEL pulse structure mimics the pulse structure of the electron beam, and the recent advances in accelerator technology have opened up the possibility of FELs operating with bursts of sub-picosecond pulses repeating at up to MHz repetition rate for macro-pulse duration from microseconds to milliseconds. Since the active medium in an FEL is a relativistic electron beam propagating through the undulator, there is no limit on the power handling capacity of the gain medium. With an electron beam to FEL conversion efficiency typically $< 1\%$, peak power ~ few MW is easily achievable in these FELs with a peak current of few 10s of Ampere in an electron beam of energy 10–50 MeV.

An Infra-Red Free Electron Laser (IR-FEL) has been built and successfully commissioned at the Raja Ramanna Centre for Advanced Technology (RRCAT), Indore, which delivers tunable radiation in the 12.5–50 μm wavelength region to a user facility [1]. This is the first and only FEL in the country to achieve lasing. The next section discusses salient features of the IR-FEL, which is followed by a discussion of the existing and planned use stations at the facility. The concluding section discusses possible future directions for the FEL activity.

2 The Infra-Red Free Electron Laser (IR-FEL) at RRCAT

The IR-FEL at RRCAT is designed to lase in the 12.5–50 μm wavelength range with $>$ 2 MW peak out-coupled IR power in 10 ps pulses, thus complementing the Indus Synchrotron Radiation Sources (SRS) at RRCAT serving users primarily in the short wavelength regions from UV to hard X-rays. Though Indus-1 SRS also delivers IR radiation for user experiments, it is designed for a critical wavelength of 61 Angstrom resulting in the photon flux in the IR region being relatively low as compared to that from the IR-FEL. The IR-FEL employs a Linac based injector system capable of delivering an electron beam with energy variable from 15–25 MeV, and a 2.5 m long pure permanent magnet undulator with an undulator parameter variable from 0.5–1.25 by varying the undulator gap. Continuous tunability of wavelength is achieved by a combination of varying the undulator gap and the electron beam energy. It is designed to operate in an oscillator configuration, and the undulator is mounted inside a 5.04 m long optical cavity with a hole in the downstream mirror for out-coupling of laser radiation. Lasing in this configuration evolves from shot noise through the repeated interaction of an optical pulse bouncing back and forth inside the optical cavity with successive fresh electron micro-bunches injected into the undulator. The IR-FEL takes around 80 round-trips to achieve saturation of lasing, after which almost uniform power output is available for user experiments for the remaining portion of the electron beam macro-pulse, typically for around 4–6 microsecond. The macro-pulses repeat at a Pulse Repetition Rate (PRR) variable from 1–4 Hz.

Fig. 1. The IR-FEL setup inside a radiation shielded vault.

Figure 1 shows a picture of the IR-FEL setup inside a radiation shielded area. The injector Linac system starting with an electron gun is at the right bottom, and the undulator and the FEL optical cavity are at the left top in the picture. Figure 2 shows the wavelength tuning curve achieved at the IR-FEL facility, with the full design range of 12.5–50 μm covered with two electron beam energies. Figure 3 shows the typical stability of out-coupled IR power from the FEL over an 8-h long operation cycle with a measured σ ~ 0.18 mW and mean ~ 4.45 mW over a 6-h period. The sampled power is a fraction (typically ~ 15%) of the CW average power transported to the user stations. The peak power achieved in these experiments typically exceeds 3 MW in 10 ps pulses at the user station.

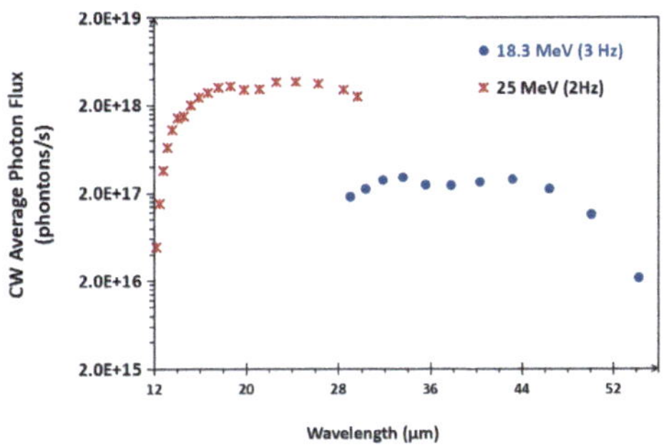

Fig. 2. Wavelength tuning curves for the IR-FEL.

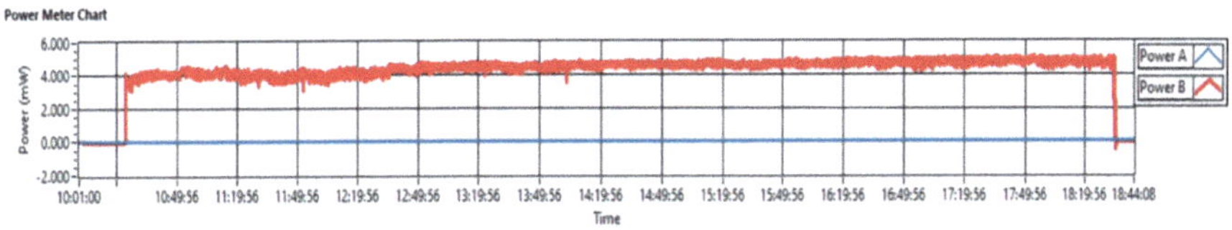

Fig. 3. Typical stability of the out-coupled IR power at a wavelength of 20 μm over an eight- hour long operation cycle.

3 The IR-FEL Based User Facility

FEL radiation from inside the radiation shielded vault is transported to a user facility over a distance > 50 m through a dry Nitrogen flushed optical beam transport line with reflecting optics. Figure 4 shows the typical pulse structure of the IR-FEL radiation, which comes in bursts of MW peak power micro-pulses of ~ 10 ps width at a frequency of 29.75 MHz or 59.5 MHz for a macro-pulse duration ~ 4–6 μs. The PRR of these macro-pulses can be varied from 1–4 Hz resulting in an increase in the CW average out-coupled power through increase in duty cycle of operation of the machine. The peak power, however, stays the same.

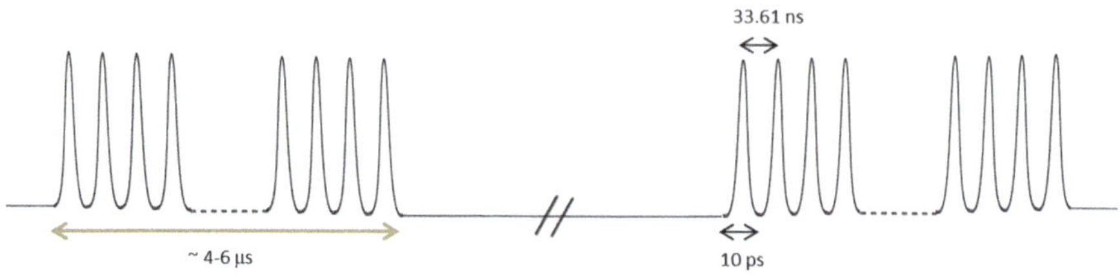

Fig. 4. Typical pulse structure of the IR-FEL radiation.

Measurements at the user port have demonstrated typical CW average power > 5 mW, more than 3 MW peak power, and 30–50 μJ energy per micro-pulse. The typical measured spectral width of FEL radiation at the user station is ~ 1.4% at a central wavelength of ~ 22 μm. Figure 5 shows a typical measured transverse profile of the IR-FEL radiation at a user station (Port 4), an interferogram obtained using a Michelson Interferometer developed at the port, and the measured spectral profile of the IR-FEL radiation.

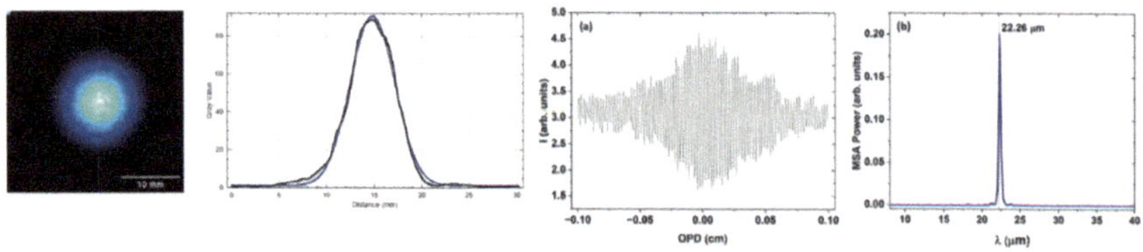

(a) (b) (c) (d)

Fig. 5. [a & b] Transverse optical mode profile of the IR-FEL radiation at a user station, [c] typical interferogram obtained using a Michelson Interferometer at Port 4 (OPD is optical path difference), and [d] the measured spectrum of the IR-FEL radiation

The IR-FEL based user facility is designed with six user ports, out of which two have been developed and commissioned for user experiments [2]. Figure 6 [Left] shows a picture of the user station at Port 3 for frequency domain spectroscopy of materials using the IR-FEL light, with provision for a sample environment with high magnetic field (7 T) and low temperature (4 K). Figure 6 [Right] shows a picture of Port 4 with an experimental station for wavelength determination using a Michelson Interferometer, wavelength selective and simultaneous reflection and transmission imaging, and for pulse slicing and polarization control for future time-domain experiments. Experiments for ambient temperature transmission/absorption characterization of samples are also performed at this port, and provision is being made for experiments on photo-current measurement at this port.

Port 1 is presently employed for FEL diagnostics, and is earmarked for the future development of stations for Biology and Chemistry experiments. Port 6 is being developed in collaboration with the UGC-DAE Consortium for Scientific Research (CSR), Indore for IR-FEL based time domain spectroscopy with a femtosecond laser system being coupled to the IR-FEL radiation at this port. Port 5 will have a station for FEL pump-THz probe studies, and plans for Port 2 will evolve in the future based on user requirements.

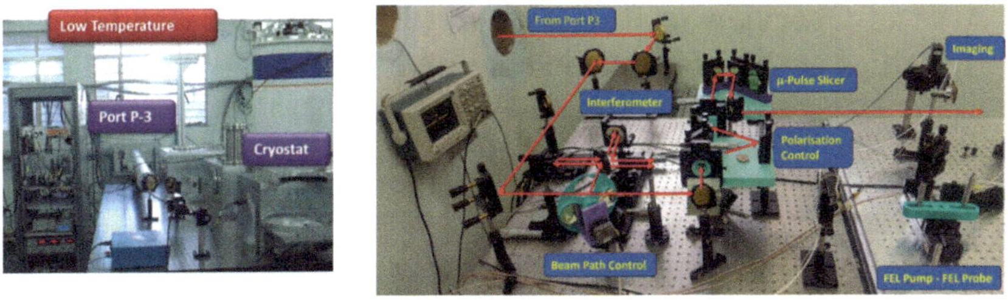

Fig. 6. [Left] Experimental station for IR-FEL based frequency domain spectroscopy of materials under low temperature (4 K) and high magnetic field (7 T) sample environment; [Right] Experimental station with a Michelson Interferometer for wavelength selective imaging, with provision for polarization control and pulse slicing for future experiments.

4 Possible Directions for the Future: An X-ray FEL (XFEL) Test Facility

Worldwide, six hard X-ray FELs have lased and one more is expected to lase in 2026. All these FELs started with the Self Amplified Spontaneous Emission (SASE) mode, and subsequently employed seeding schemes for better experimental conditions with lower timing jitter and better spectral quality [3]. Such facilities require huge investments in terms of cost, real estate and equipment. Many other short wavelength FELs are providing short wavelength radiation in the soft x-ray and VUV region by employing alternate schemes like seeding and harmonic generation, which lowers the electron beam energy and undulator length requirements for generating similar wavelengths. This makes the facility relatively smaller and less cost intensive. Both kinds of short wavelength FELs are enabling cutting-edge user experiments in different fields of science and technology. Such machines require the generation, manipulation and diagnostics of ultra-short and ultra-intense electron beams of very good quality, imposing very stringent requirements for associated sub-systems like the Low-level Radiofrequency (LLRF) and timing systems, high power microwave systems, beam transport line sub-systems. A natural direction for the FEL activity, therefore, is to embark on a path towards building a Soft X-ray FEL employing an electron beam with properties similar to those required for the ultimate goal of building a hard XFEL. This would serve as an XFEL Test Facility and may be accomplished in a phased manner, with the first phase being the construction of an Ultraviolet (UV) FEL employing a 250 MeV electron beam, to be followed in subsequent phases by a 1–1.5 GeV machine on the lines of seeded FELs built worldwide to complement the hard XFELs. All state-of-the-art technologies required for building XFELs, like bunch compression to achieve femtosecond electron bunches, SASE using long undulator sections, seeding and harmonic generation schemes, precise timing and RF and microwave systems would be demonstrated in the first phase that would also deliver a UV-FEL based user facility. The space requirement for the first phase would be relatively modest and it can be accommodated within a shielded area ~ 60 m long for the FEL. Subsequent phases will build on the technologies of the previous phase, ultimately leading to the development of an XFEL Test Facility.

5 A Brief Overview of Research and Development (R&D) Activities on Accelerator Science and Technology Being Pursued in the Department of Atomic Energy (DAE)

Institutes under the DAE play a leading role in R&D related to accelerator science and technology, and are home to most of the major accelerator facilities in the country. Accelerator R&D in the DAE is driven primarily by the goals set under the Vertical-3 of the Amrit Kal Vision Document of the Department, which has a very good overlap with the Mega Science Vision Document 2035 on Accelerator S&T and applications. The activities pursued in the Department can primarily be categorized as follows: (i) Accelerator based facilities for scientific research, (ii) Accelerator based facilities for medical and industrial applications, and (iii) Accelerator R&D related to the High Intensity Proton Accelerator (HIPA) program.

The Variable Energy Cyclotron Centre (VECC) in Kolkata is home to three operating cyclotrons that deliver beams for scientific research and for the commercial production of radio-isotopes. The room temperature cyclotron delivers beam of alpha particles (28–50 MeV), protons (7–12.5 MeV), and ions of Nitrogen (105–140 MeV), Oxygen (116–160 MeV) and Neon (145–192 MeV), which are used for nuclear physics research and radiation damage studies. It is also used for Radioactive Ion Beam (RIB) technology demonstration facility testing. The super-conducting cyclotron delivers beams of Nitrogen (250–415 MeV), Oxygen (309–416 MeV), Neon (360–490 MeV), Argon (190–340 MeV) and Krypton (415–650 MeV) ions for nuclear physics research and for radiation damage studies. The Medical Cyclotron delivers proton beams of 15–30 MeV with 500 mA current, and it used by the Board of Radiation and Isotope Technology (BRIT), DAE for the production of radio-pharmaceuticals like Flouro Deoxy Glucose (FDG), Sodium Flouride, Thallus Chloride - TI-201 for Single-Photon Emission Computed Tomography (SPECT), and 68 Gallium Prostate Specific Membrane Antigen (PSMA). The VECC campus also houses a technology demonstrator RIB facility, which has delivered radioactive beams of ^{14}O, ^{42}K, ^{43}K, ^{111}In, ^{11}C in addition to stable beams of Carbon, Oxygen, Nitrogen, Argon, Nickel, Silver, Zinc, Iron and Boron with beam energy from 10–715 keV/u. An indigenous medical cyclotron is under development at VECC to deliver 18 MeV proton beam with 50 μA current, with an aim to transfer the technology to an Indian industry. A proton cyclotron (50 MeV, 200 μA) is also being developed as a primary accelerator for the project 'Applied and Nuclear Research with Rare Isotope Beams' (ANURIB) coming up at VECC's Rajarhat Campus. A Radio Frequency Quadrupole and linear accelerator structures for the ANURIB project have already been built and tested, and are ready for deployment.

The Bhabha Atomic Research Centre (BARC), Mumbai is working on a three-stage plan for the development of High Intensity Proton Accelerators in the country. In the first stage, called as the Low Energy High Intensity Proton Accelerator (LEHIPA), proton beam acceleration up to 20 MeV has been successfully achieved employing an Electron Cyclotron Resonance (ECR) Ion Source, RFQs and Drift Tube Linacs (DTL) developed indigenously. This is the normal conducting front-end for the HIPA program. The proton beam from LEHIPA will be used for nuclear physics and materials science research, and for studies on the generation of radioisotopes. The next planned phases are the Medium Energy High Intensity Proton Accelerator (MEHIPA) Phase I and II, which aim at delivering 200 MeV proton beam in the year 2035. These two phases will employ superconducting technology for which the science and technology of building spoke resonators cavities, cryomodules and associated cryogenic technologies, RF technologies, and facilities are under development. Following the MEHIPA Phase I and II, the High Energy High Intensity Proton Accelerator (HEHIPA) Phase I and II will be taken up ultimately leading to the development of a 1 GeV machine by the year 2042. The high energy Linac will employ superconducting elliptical cavities and will require the simultaneous development of significant cryogenic and RF infrastructure. It is worth mentioning here that as a part of India's contribution to the IIFC program, the learning curve associated with different technologies required for the MEHIPA Phase I and II and also for the HEHIPA Phase I and II has already been ascended, and deliverables like elliptical cavities, cryomodules, spoke resonators, etc. under the collaboration have already been developed and delivered to FERMILAB by DAE Institutes.

BARC also pursues R&D activity on electron Linacs, primarily aimed at technology development for industrial applications. An indigenous dual energy cargo scanner has been developed with provision to operate at dual energies of 6 MeV/4 MeV with $\leq 2\%$ dose stability (σ) and a through put of 80 TPH. Work is currently underway to ultimately achieve a through put of 150 TPH following which technology would be transferred to an Indian industry. A 6 MeV X-band Linac has been developed and successfully tested at the Electron Beam Centre (EBC), and it is being incubated with an Indian industry to ultimately build a cancer therapy machine employing this technology. R&D activity is also underway on the development of a 30 MeV normal conducting electron Linac for RIB generation, which will deliver 19 kW electron beam power. A 15 MeV prototype is currently under testing at EBC. An essential technology for all electron Linac projects is a high power klystron, which will be required in large numbers to meet the national needs for research, industrial and medical electron Linacs in the future. BARC is coordinating an effort towards the indigenous development of a 6 MW peak power and 24 kW average power klystron at 2856 MHz.

The Tata Institute for Fundamental Research (TIFR), Mumbai houses the Pelletron-Linac Facility, which has been built and is operated jointly by BARC and TIFR, and is primarily used for research applications. The Pelletron runs as a stand-alone facility, as well as an injector to the Linac. Around 50 user experiments are conducted at the facility each year, and the beams accelerated through the Linac include ^{7}Li, $^{10,11}B$, ^{12}C, $^{16,18}O$, ^{19}F, $^{28,30}Si$, $^{32,34}S$, ^{35}Cl, ^{37}Cl. The facility has produced ~ 830 publications in refereed journals, including ~ 35 Letters, and has contributed to ~ 160 Ph.D. theses.

RRCAT, Indore is primarily a light sources laboratory with the two Synchrotron Radiation Sources (SRS) Indus 1 and 2 operated round-the-clock as user facilities supporting research by users from the academia as well as industry. Indus 1 is a second generation SRS with a 450 MeV, 125 mA electron beam, and it delivers broadband bending magnet radiation with a critical wavelength of 61 Angstrom to users at 6 operational beamlines. Indus 2 is a third generation SRS with a critical wavelength of 1.98 Angstrom, and it delivers radiation from bending magnets or Insertion Devices at 17 user beamlines. The Indus SRS have a user base of more than 140 academic institutes and more than 20 industries, and more than 1000 user experiments are conducted every year at the Indus beamlines in the fields of material science, healthcare, medicine, agriculture, environment, energy, manufacturing, semiconductors, chemicals and petrochemicals. More than 1000 research publications and more than 200 Ph.D. theses have come out of the research done using these facilities. With a growing demand from users for a fourth generation light source to perform contemporary experiments, detailed studies have been initiated for the design of a proposed Indus 3 SRS and its sub-systems. This machine is proposed to be operated at 6 GeV with 200 mA beam current, and will be operated in the top-up mode. It is designed to have a beam emittance < 150 pm.rad, which is 300 times better than that of Indus 2, and a peak Brilliance six orders of magnitude higher than that of Indus 2. Thirteen beamlines are planned for the first phase of the machine. Considering the above, the specifications of different sub-systems for the proposed Indus 3 are much more stringent than those for Indus 2, and activities have been initiated on the design and development of such sub-systems in the technology development phase.

The IR-FEL at RRCAT is another accelerator based light source that complements the Indus SRS in the long wavelength region. It delivers more than two orders of magnitude higher Brilliance than Indus 1 in the 12.5–50 μm wavlength range, thereby facilitating research in areas requiring low photon energies. It is now operating as a user facility facilitating research by users from within as well as outside RRCAT.

RRCAT also has a focused activity on the development and operation of a 10 MeV electron Linac based facility for the commercial sterilization of medical products. A 10 MeV, 6 kW Linac based system with a processing capacity of 200 kg/hour for a minimum dose of 25 kGy has successfully sterilized more than 1.83 crore medical devices, which are supplied pan India and also to more than 35 countries.

RRCAT has also developed elaborate infrastructure for the development, testing and qualification of SCRF cavities, which are being used for the development and qualification of deliverables under IIFC, and which will support the activities related to the MEHIPA Phase I and II at BARC.

The Nuclear Fuels Complex (NFC), Hyderabad, in collaboration with RRCAT has indigenously developed Niobium material with Residual Resistivity Ratio (RRR) > 300, which is essential for building superconducting cavities and which will be required in large quantity for the future projects of the DAE.

6 Summary

The IR-FEL at RRCAT, the first FEL in the country to lase, has successfully achieved its design goals, and it now serves as a user facility delivering intense, short-pulsed IR radiation with full wavelength tunability from 12.5–50 μm wavelength for scientific research. The facility presently facilitates experiments like frequency domain IR spectroscopy, wavelength resolved imaging in reflection transmission modes, study of nonlinear phenomena, etc. With the addition of a facility for time domain spectroscopy, currently under development, the IR-FEL based user facility will facilitate frequency as well as time-domain studies of samples from different areas of science. With the successful demonstration of FEL science and technology in the long wavelength region, a logical and possible future direction for the FEL activity is to build an XFEL test facility in a phased manner, with the first phase being the development of a UV-FEL to demonstrate state-of-the-art technologies essential for a future XFEL.

In line with the Amrit Kaal Vision Document of the DAE, and also in line with the National Mega Science Vision Document - 2035 on Accelerator S&T and Applications, institutes under the DAE play a leading role in R&D related to accelerator science and technology, and are home to most of the major accelerator facilities in the country. Different DAE institutes have also embarked upon ambitious projects to build new accelerators and related technologies to support frontline scientific research, for industrial and medical applications, and to meet the requirements of a HIPA for Accelerator Driven Systems in the future.

References

1. Chandran, S., et al.: The IR-FEL facility at RRCAT: commissioning experiments and first saturation of lasing at 28 μm wavelength. Nucl. Inst. Methods Phys. Res. A **1003** 165321 (2021)
2. Chattopadhyay, M.K., et al.: Establishment of a free electron laser-based user facility for IR-THz spectroscopy at RRCAT'. Curr. Sci. **128**(7) (2025)
3. Huang, Z.: Review of advanced schemes for free electron lasers. In: IPAC 2024. Tenesse, USA, May 19–24 (2024)

Seeing the Unseen: Detectors for Nuclear and High Energy Physics Research

Tilak Kumar Ghosh[1,2](✉)

[1] Variable Energy Cyclotron Centre, 1/AF Bidhannagar, Kolkata 700064, India
[2] Homi Bhabha National Institute, Training School Complex, Anushakti Nagar, Mumbai 400094, India
tilak@vecc.gov.in

Abstract. Radiation detectors are central to nuclear and high energy physics, enabling the measurement of particles and radiations that cannot be observed directly. Over the past several decades, laboratories under the Department of Atomic Energy (DAE), India, have developed strong indigenous capabilities in detector design, fabrication, characterization, and system integration. These efforts encompass charged particle detectors, gas-based detectors, neutron and scintillation detectors, and high-resolution γ detectors, supporting experiments at national accelerator facilities as well as major international collaborations. This article presents a concise overview of selected detector developments with emphasis on technological capability, experimental infrastructure, and collaborative effort rather than specific physics results. Several of these detectors, developed primarily for basic research, also show promise for applications beyond fundamental science.

Keywords: Radiation detectors · detector R&D in India · societal applications

1 Introduction

Detectors play an indispensable role in experimental physics. This phrase "seeing the unseen" captures the essence of what we do in nuclear and high-energy physics. We study objects and processes that are fundamentally invisible to our eyes - yet they govern the structure of matter and the universe. Detectors converts invisible particles and radiations into measurable signals. While accelerators act as microscopes to probe the microscopic world made of nucleons and its constituents, detectors serve as the "eyes" that help to understand it.

Dedicated to Dr. Sumit Som, former Director, VECC, on his 60th birthday, in recognition of his support for indigenous instrumentation and accelerator-based nuclear physics research in India.

G. B. Mohanty and V. S. Prabhudesai (Eds.): DAEC 2026, SPPHY 368, pp. 200–204, 2026.
https://doi.org/10.1007/978-981-92-2948-2_26

India's nuclear and high energy physics programmes, largely coordinated through DAE laboratories, have consistently emphasized indigenous detector development. Indian laboratories have progressed from being users of imported systems to designers and builders of sophisticated detectors. These efforts support experiments at national facilities such as VECC, TIFR, BARC, IUAC, and RRCAT, and contribute significantly to international programmes at CERN, FAIR, and RHIC.

This article provides a selective overview of detector developments presented at the DAE Conclave. Rather than offering a comprehensive review, it highlights representative examples that illustrate the breadth, maturity, and coherence of detector R&D within the DAE ecosystem. The work summarized here reflects the collective contributions of many researchers and institutions across the country.

2 Principles of Radiation Detection

Radiation detectors operate by converting energy deposited by incident radiation into electrical signals. Depending on the interaction mechanism and detection medium, this conversion occurs through ionization, excitation, or charge carrier generation. Broadly, three classes of detectors dominate nuclear and high energy physics experiments: semiconductor detectors, gas-filled detectors, and scintillation detectors.

Semiconductor detectors provide excellent energy resolution through the collection of electronhole pairs, gas detectors offer large-area coverage with good timing and position sensitivity, and scintillation detectors provide fast response and high efficiency through light emission detected by photomultipliers or silicon photomultipliers. These detector technologies form the foundation of experimental setups at Indian accelerator facilities, with many systems being developed indigenously.

3 Charged Particle and Gas-Based Detectors

Charged particle detection is central to nuclear reaction studies. A flagship example is the Charged Particle Detector Array for Kinematic Reconstruction and Analysis (ChAKRA) [1], a high-granularity 4π silicon and CsI (Tl) detector array designed and developed at VECC for complete charged particle spectroscopy. ChAKRA enables detailed investigations [2] of reaction dynamics, clustering phenomena, transfer reactions, breakup processes, and astrophysically relevant states such as the Hoyle state in carbon. Its modular and flexible geometry allows adaptation to diverse experimental requirements, and parts of the array are now operational at the K500 Superconducting Cyclotron in Kolkata. Recently, particle identification has been demonstrated at BARC using digital pulse shape discrimination with a single silicon detector, eliminating the need for conventional ΔE–E telescopes. This approach enables compact detector setups and is particularly useful for low-energy nuclear reaction studies.

Gas detectors continue to play a vital role, especially in heavy-ion reactions and fission studies. Multi-wire proportional counters (MWPCs) developed [2] and deployed at several Indian laboratories provide precise timing, position, and energy-loss information and remain workhorse for fission research. Building on MWPC expertise, hybrid detector systems combining gas detectors with silicon detectors have been developed to provide simultaneous measurements of time, position, and energy, enhancing sensitivity to reaction mechanisms relevant to the synthesis of superheavy elements.

4 Neutron, Scintillator and Gamma Detectors

Neutron detection is essential for studying reaction dynamics, fission processes, and nuclear structure. At VECC, a neutron time-of-flight detector array based on liquid scintillators has been developed [2] and deployed at the K500 beam hall. Detector fabrication has been fully developed in-house. Earlier such detectors were used to be imported, but Indian laboratories are now sufficiently advanced to fabricate such detectors. VECC has also fabricated neutron detectors to be used in the MONSTER array at FAIR, Germany. BARC group have deployed plastic and liquid scintillator arrays at TIFR for fast neutron spectroscopy, supporting studies of nuclear level density, transfer reactions, and fission dynamics.

Scintillator development remains an active area of R&D. Deuterated liquid scintillators, developed through collaborations involving BARC, HWB, and TIFR, show improved pulse-height resolution and enhanced safety characteristics compared to conventional scintillators. The establishment of an indigenous plastic scintillator fabrication facility at the Cosmic Ray Laboratory, TIFR–DAE, Ooty, represents a significant step toward self-reliance in scintillator detector development in India.

High-resolution gamma-ray spectroscopy remains a cornerstone of nuclear structure research. The Indian National Gamma Array (INGA), comprising Compton-suppressed Clover HPGe detectors, operates as a national multi-institutional facility across accelerator centers. Complementary detector arrays at VECC (e.g.; VENUS, VENTURE) and BARC (e.g.; DURGA) support fast timing, angular correlation, and decay spectroscopy measurements. Indian expertise, led by the TIFR group, also contributes to the DEGAS detection system for the NuSTAR program at FAIR. The 4π spin spectrometer developed at TIFR enables evaporation-residue-gated spin measurements in nuclear reaction. The Large Area Modular BaF_2 Detector Array (LAMBDA) with 162 detectors is developed by VECC for high-energy gamma ray detection.

5 Detectors for High Energy and Astrophysics

India has also made significant hardware and instrumentation contributions to high energy physics experiments at CERN and RHIC. Notably, India plays a key role in the ALICE FoCal upgrade through the indigenous design and fabrication of advanced silicon detectors in collaboration with Indian industry. In astroparticle physics, the Major Atmospheric Cherenkov Experiment (MACE) telescope, designed by BARC and built by ECIL, reflects India's growing capability in large-scale scientific instrumentation.

6 Detectors Beyond Fundamental Research

Although developed primarily for basic research, many detector technologies within the DAE ecosystem have clear relevance for societal and strategic applications. The scintillator-based ISMRAN (Indian Scintillator Matrix for Reactor AntiNeutrino), developed for antineutrino experiments, provides a technological basis for non-intrusive reactor monitoring. Gas Electron Multiplier (GEM) detectors developed for the FAIR program show strong potential for high-resolution medical imaging; high-resolution X-ray imaging using GEM detectors has been demonstrated at VECC. Resistive Plate Chambers (RPCs) developed at SINP and VECC for the Indian Neutrino Observatory offer promising applications in muon tomography, particularly for the detection of high-Z materials (e.g.; Thorium, uranium) relevant to homeland security. In addition, position-sensitive scintillator detectors [3] developed at TIFR show potential for gamma-ray imaging applications. Together, these examples illustrate how sustained investment in detector R&D for fundamental science creates a strong foundation for broader technological and societal impact.

7 Summary and Outlook

Detector development is a core strength of the DAE research ecosystem. Indian laboratories have demonstrated the capability to design, fabricate, and operate sophisticated detector systems spanning nuclear, particle, and astroparticle physics. These detectors support national research programmes, contribute significantly to international collaborations, and increasingly enable applications beyond fundamental science. Continued emphasis on indigenous detector development, supported by inter-institutional collaboration and industry partnerships, will remain central to advancing scientific capability while supporting India's vision of self-reliance in advanced instrumentation.

8 Disclaimer and Acknowledgments

This article is based on contributions from many researchers and laboratories across India; any omissions are due to space limitations. The author thanks colleagues who shared the material and insights. AI-assisted language tools were used to support editing and refinement of the text.

References

1. Kundu, S., et al.: ChAKRA: the high resolution charged particle detector array at VECC. Nucl. Inst. Methods A **943**, 162411 (2019)
2. Bhattacharya, S., Bhattacharya, C., Das, P., Mukherjee, G., Ghosh, T.K.: Experimental nuclear physics research using the cyclotron at VECC. Eur. Phys. J. A **54**(9), 1–38 (2018). https://doi.org/10.1140/epja/i2018-12586-2
3. Das, B.: Development of position sensitive scintillator detector. Nucl. Inst. Methods Phys. Res. A **1048**, 167928 (2023)

Development of Special Materials and Components for Nuclear and Non-Nuclear Applications at the Nuclear Fuel Complex

G. N. Ganesha(✉)

Nuclear Fuel Complex, Hyderabad 500062, India
ganesh@nfc.gov.in

Abstract. The talk covers manufacturing activities for seamless tubes of different grades, special materials, components & assemblies developed for nuclear & non-nuclear applications.

Keywords: Tubing · Assemblies · Nuclear applications · Non-nuclear applications · Facilities

1 Introduction

The talk covers manufacturing activities for seamless tubes of different grades, special materials, components & assemblies developed for nuclear & non-nuclear applications such as:

In-core and out-of-core structural components for thermal & fast breeder nuclear reactors.

Titanium & Nickel base alloy tubing for aerospace applications.

Tubing & components of different grades for strategic & defense applications.

Tubing of different types of Nickel base super-alloy for thermal power and reprocessing plants.

Process development and manufacturing of high RRR Nb sheet for SCRF cavity application.

NFC continuously develops advanced manufacturing process routes for new alloy tubes & components required for nuclear reactors, defense & space applications. The manufacturing flow sheet for typical seamless tubes include the following steps: omponents required for nuclear reactors, defense & space applications. The manufacturing flow sheet for typical seamless tubes include the steps shown in Fig. 1.

2 Tubing & Assemblies for Nuclear Applications

NFC is involved in manufacturing of various type of fuel assemblies and structural components for thermal reactors like PHWR and BWR (Fig. 2). These include Zr-2 and Zr-4 clad tubes and structural components: Zr-2.5%Nb Pressure Tube, Zr-4 Calandria

G. B. Mohanty and V. S. Prabhudesai (Eds.): DAEC 2026, SPPHY 368, pp. 205–208, 2026.
https://doi.org/10.1007/978-981-92-2948-2_27

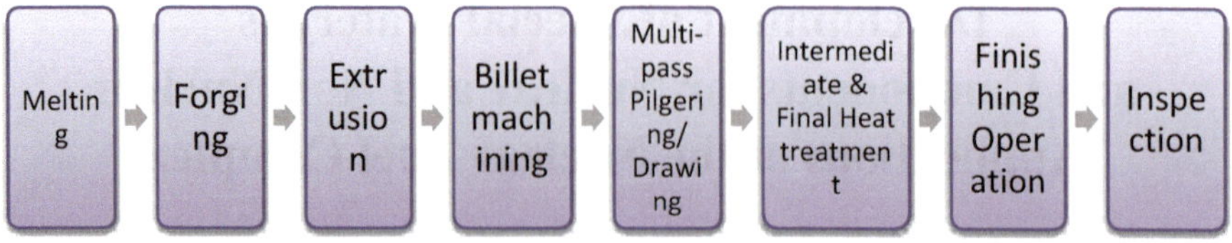

Fig. 1. Steps involved in manufacturing flow sheet for typical seamless tubes.

Tubes, Zr-2.5Nb-0.5Cu Garter springs, Zr-4 square channels. NFC has joined the few elite group of companies which have mastered the technology to manufacture UNS No. 8800 tubes with 26 to 28 m length, U-bend in 72 radii, 91 to 1014 mm CLR for steam generator of PHWR. Annular Cobalt Capsule (Zr-4) tubes required for Adjuster Rods in DHRUVA reactor are also manufactured in NFC.

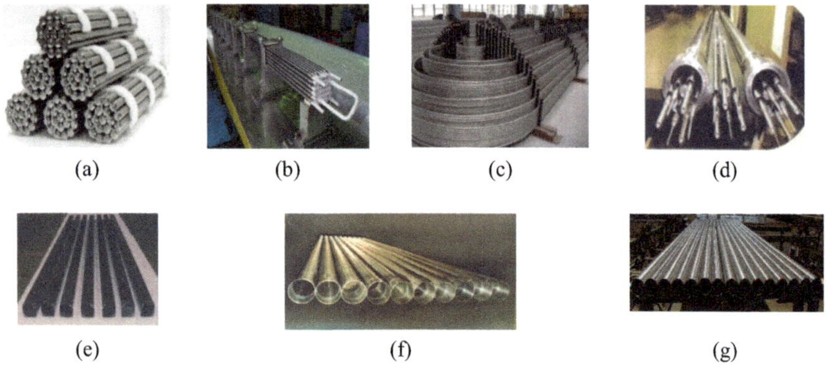

Fig. 2. (**a**) PHWR fuel bundle, (**b**) BWR fuel bundle, (**c**) U bend SG tubes for PHWR, (**d**) reactivity mechanism assemblies for PHWR, (**e**) square channels, (**f**) calandria tubes, and (**g**) pressure tubes.

The process route has been modified for manufacturing of pressure tubes to include radial forging followed by extrusion and single pass pilgering. The modification resulted in uniform microstructure and uniform mechanical properties along length of the tube. This modification also resulted in superior texture (higher F_T–F_R) and more favorable grain aspect ratio.

NFC is also contributing in the manufacturing of various grades of alloys for fast breeder reactor applications. These include D9/D9I clad tubes (Fig. 3), Zr-T91 double clad tubes. In addition to this, high burn-up clad tubes (up to 200 dpa) made of 9Cr & 18Cr ODS have also been developed and successfully supplied. These tubes offer best combination of void swelling resistance of ferritic-martensitic steels with high-temperature creep resistance comparable to austenitic steels. Steam generator tubes of T91. A unique process has been developed to manufacture a hexcan from an ingoing solution annealed hexcan to impart uniform retained cod work across cross section.

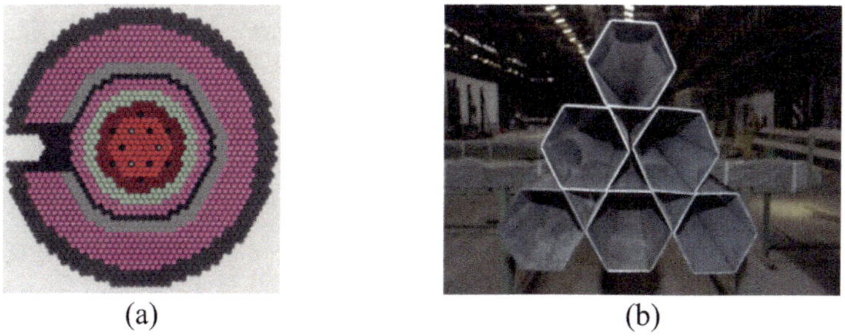

Fig. 3. (a) D9/D9I clad tubes for PFBRs and (b) hexcans

2.1 Tubing of Different Types of Nickel Base Super-Alloy for Thermal Power and Reprocessing Plants

Hot extrusion and cold pilgering parameters were established in NFC for Ni based super alloy tubes conforming to ASTM B983, UNS N06617 tubing and 304HCu tubing (Fig. 4). These tubes were successfully developed in collaboration with Midhani and IGCAR for boiler tubes for Advance Ultra Super Critical application.

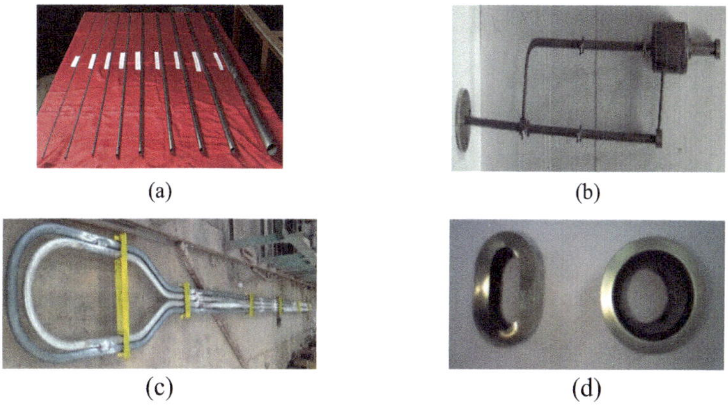

Fig. 4. (a) Ni based super alloys UNS N06600 and UNS N06690 tubing, (b) Nb-1Zr-0.1C test loop for CHTR, (c) UNS N06617 test loop, and (d) 304 HCu Tubes

3 Tubing & Assemblies for Non-Nuclear Applications

3.1 Space Applications

Nickel base alloy UNS N04400 for semi cryogenic engine (Fig. 5), Titanium Half alloy for structural material for Gaganyan project, and Aluminum Alloy AA 2219 for propellant feed lines of liquid oxygen & liquid hydrogen were successfully manufactured and supplied to ISRO. These tubing components are being used in various launch vehicles and space applications.

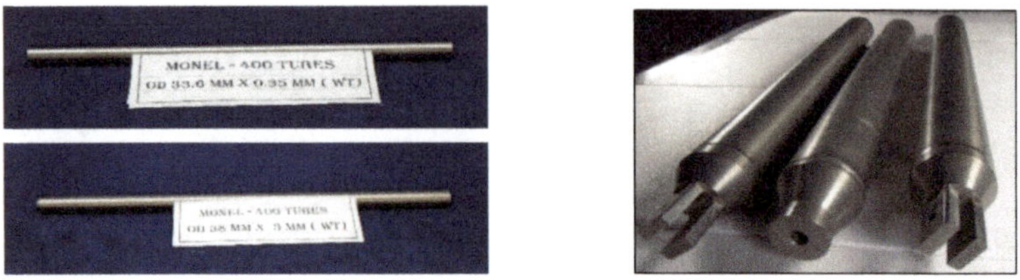

(a) (b)

Fig. 5. (**a**) UNS N04400 tubes and (**b**) Ti-3A1–2.5 V alloy truss rod assembly

3.2 Other DAE Application

In addition to tubular products, NFC also manufactures several other products such as RRR Niobium Sheet, Micron Size Zirconium powder. RRR Niobium is primary material for Super Conducting Radio Frequency (SCRF) Cavities. At present, only with 4 countries in the world have mastered the Technology of RRR Nb. Towards achieving "Aatmanirbhar Bharat", NFC successfully has developed the technology for the manufacture of RRR Nb ingot and sheet by successfully achieving highest ever Residual Resistivity Ratio (RRR) value of 334 for the first time in the country. 2 sheets of RRR Nb have been supplied to RRCAT for the manufacture of SCRF cavities (Fig. 6).

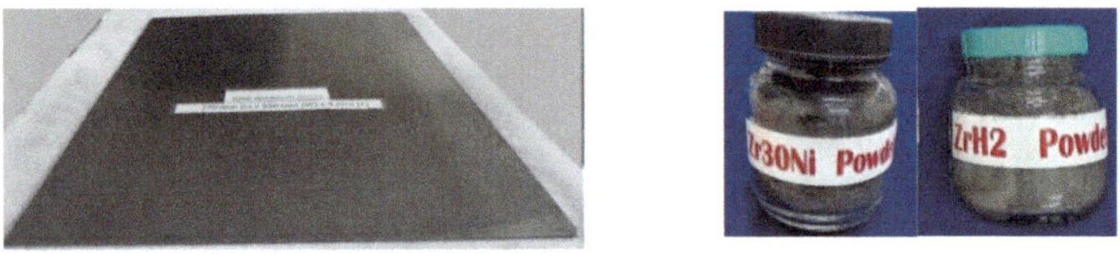

(a) (b)

Fig. 6. (**a**) RRR Nb sheet, and (**b**) Zr metal powder

The above works are collaborative in nature and has been carried out on with various units of DAE, PSUs, other organizations belonging to Ministry of Defense and Department of Space. NFC gratefully acknowledges the contributions made in the development of various alloys, special materials and components.

4 Manufacturing Facilities

NFC is having capability of manufacturing tubes of different materials such as Zr, Nb, Ti, Fe & Ni based alloys. Tubes of outer diameter ranging from 4 mm to 150 mm & wall thickness ranging from 0.25 mm to 15 mm are successfully manufactured and supplied to various unit of DAE, Defense & space. The state of art facilities available for manufacturing of the tubes are shown in Fig. 7.

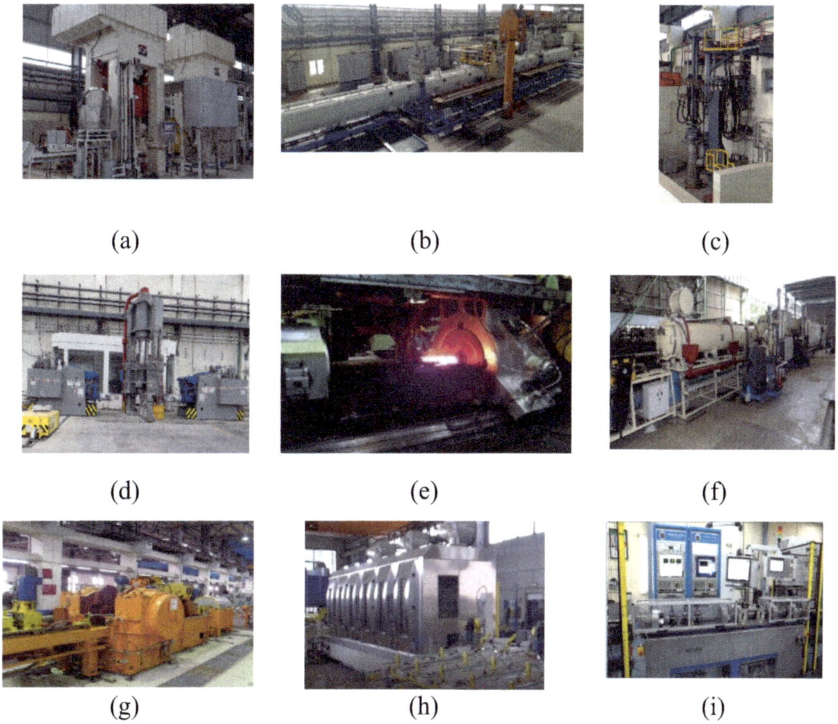

Fig. 7. (**a**) 2000 T compaction press, (**b**) electron beam welding unit, (**c**) 350 dia. Vacuum arc melting furnace, (**d**) 1600 T forge press, (**e**) 3780 Te horizontal extrusion press, (**f**) 9 m vacuum annealing furnace, (**g**) 65 mm cold reducing pilger mill, (**h**) Tube polishing machine, and (**i**) Immersion Tube Ultrasonic Testing Machine.

R&D Activities Related to Fusion Technologies at Institute for Plasma Research

Paritosh Chaudhuri(✉)

Institute for Plasma Research, Gandhinagar 382428, India
paritosh@ipr.res.in

Abstract. The Institute for Plasma Research (IPR) was established in 1986 with the primary objective of promoting fundamental and applied research in plasma physics and fusion technology. It is an aided institute of the Department of Atomic Energy (DAE). The institute has grown significantly over the years, becoming India's leading centre for plasma science and fusion technology. This paper presents a comprehensive overview of the research and development (R&D) activities at IPR, focusing on basic plasma physics, magnetically confined plasmas, plasma technologies for industrial applications, advanced engineering systems, and international collaborations such as ITER. The paper highlights IPR's contributions to R&D challenges on plasma science and fusion technologies, technological innovation, and human resource development.

Keywords: Tokamak · Plasma · Aditya-U · SST-1 · ITER · LIGO · Plasma Facing Components · Blanket · Magnet

1 Introduction

The Institute for Plasma Research (IPR), located in Gandhinagar, Gujarat, is a premier scientific research institute under the Department of Atomic Energy (DAE) [1]. IPR is largely involved in R&D activities in the field of plasma and fusion research which includes experimental, theoretical, and computational plasma studies, development & operation of magnetically confined plasma devices called tokamak. Its interdisciplinary approach integrates physics, engineering, and materials science to address complex challenges in fusion energy and industrial applications. IPR's fusion technology programs continued to strengthen indigenous Tokamak capabilities. Experimental campaigns on ADITYA-U, SST-1 (Steady state Superconducting Tokamak) and SSST (Small Scale Spherical Tokamak) Tokamak strengthen studies in plasma confinement, control, and steady-state operation. Applications of plasma for societal and industrial applications also constitute an important component of IPR's research activities. As a constituent institute of Homi Bhabha National Institute (HBNI), IPR also engages in diverse academic activities, including PhD and PDF programs. IPR is the nodal agency for India's participation in the ITER (International Thermonuclear Experimental Reactor) [2] and also a part of the consortium for research on LIGO (Laser Interferometer Gravitational-Wave Observatory) program [3]. Strong national partnerships with universities, other

G. B. Mohanty and V. S. Prabhudesai (Eds.): DAEC 2026, SPPHY 368, pp. 210–216, 2026.
https://doi.org/10.1007/978-981-92-2948-2_28

DAE units, and industry complement international collaborations with ITER and LIGO, reflecting notable progress in mega-science participation. Committed to human resource development, IPR has vibrant programmes to train young researchers, contributing to capacity building in plasma physics and fusion technologies. IPR also has a strong scientific outreach activity to promote plasma science in schools and colleges, particularly in rural areas, fostering indigenous expertise and inspiring future scientists [4]. This helps to build indigenous capability in across the country.

All the above mentioned activities of IPR are carried out in four different campuses: the main campus located in Bhat Gandhinagar, Facilitation Centre for Industrial Plasma Technologies (FCIPT) at GIDC, Gandhinagar [5], ITER-India (near IPR) for India's contributions to the ITER project [6], and Centre of Plasma Physics (CPP-IPR) at Guwahati [7]. This paper explores the major R&D activities undertaken at IPR, including fusion research, plasma physics, industrial applications, and international collaborations. The institute's contributions toward fusion technologies and technological advancement are also discussed.

2 Tokamak Research

Indian fusion program has progressed over the years with the development and operation of small-medium sized tokamaks. Aditya is the first indigenously designed and built tokamak and started operation in 1989. Later it was upgraded Aditya-U in 2016. The Aditya-U is an advanced version of its predecessor, Aditya (circular plasmas only), enabling shaped plasma operation in single/double-null open divertor configuration. This transformation enables the exploration of more reactor-relevant plasma conditions, diagnostics, and experiments. It supported extensive experimental campaigns, validation of diagnostic components, divertor and limiter improvements, and real-time plasma control developments. The Aditya -U enables focused investigations into key physics challenges, such as runaway electron management, disruption prediction and mitigation, real-time plasma position control, wall conditioning methods, auxiliary heating, current drive capabilities, and enhancing confinement in shaped plasmas [8]. To enhance the tokamak based plasma research in India, Aditya-U has also been included as a national user facility in the I-STEM portal.

The SST-1 tokamak which is a significant step from Aditya, is a medium-sized superconducting tokamak, designed to explore the physics and engineering challenges related to long-duration plasma confinement and steady-state plasma operation, which is crucial for future commercial fusion reactors. It has superconducting Toroidal Field (TF) magnets on torus Ultra-High Vacuum (UHV) vessel for external plasma confinement with a 3.0 T magnetic field and Poloidal Field (PF) magnets for shaped elongated and triangular shaper plasma operation. SST-1 is presently under refurbishment to bring it back to designed operating conditions in the shortest possible time.

India's first Spherical Tokamak (**S**mall **S**cale **S**pherical **T**okamak: SS-ST) is indigenously designed to study physics and engineering aspects of operating an economical and compact fusion device. It was built entirely with contributions from local industries, which embodies the spirit of Make in India. The first plasma in SS-ST was achieved on 11 December 2025. The SS-ST is now fully operational with plasma experiments being carried out regularly. This will serve as a training ground for scientists and engineers and building expertise in plasma science and fusion technology.

IPR has outlined a national fusion roadmap that includes an indigenous steady-state tokamak called SST-Bharat [9], about 2/3rd of ITER size, which will need to be a test bed to all critical fusion reactor technologies and concepts that are necessary for a future electricity generating fusion machine (DEMO). This will be India's path toward sustainable and self-reliant fusion energy program with an intermediate machine between experimental fusion devices and demonstration (DEMO) fusion reactor.

3 Various R&D Activities

Over the years, IPR has developed competence in advanced fields like fusion blanket, first wall and divertor, superconducting magnets and cryogenics, ultra-high vacuum technologies, sophisticated plasma diagnostic systems, RF and neutral beam heating systems, power supply systems, computer based data acquisition and control systems, computer modelling and simulation etc.

3.1 PFC & Blanket Technologies

The Plasma Facing Components (PFC) and Blankets are the important elements in a fusion reactor as both of them receive intense neutron flux and particle fluxes. The PFC faces surface heat loads, while breeding blankets absorb neutrons, breed tritium and converting them to thermal energy. Exposure to the fusion environment presents new challenges for PFC and blanket materials. Key PFC materials include tungsten (due to high melting point), beryllium (low-Z impurity control) for PFC, and reduced-activation ferritic/martensitic (RAFM) steels are considered for structural material for blanket. This Indian RAFMS has been developed jointly by IPR, IGCAR, and MIDHANI which replaces elements like Molybdenum (Mo) and Niobium (Nb) with Tungsten (W) and Tantalum (Ta) which have high residual activity on neutron exposure. Lithium-based compounds are used as the functional material (to breed tritium in the blanket). Several experimental facilities have been developed at IPR for the measurement and characterizations of these materials for their qualification. IPR has also established several facilities which are the first of a kind in India, to validate fusion components.

Experimental Helium Cooling Loop (EHCL), a high-pressure (8.0 MPa), high-temperature (300–450 °C) system for testing blanket mock-ups [10].

High Heat Flux Test Facility (HHFTF), uses an electron beam as a heat source to test plasma-facing components at 10 MW/m^2 heat flux [11].

Neutron Irradiation Facility (NIIF), an accelerator-based 14 MeV D-T neutron source used to study radiation damage on fusion materials and electronics [12].

Lead-lithium Loops, to perform various experiments like heat transfer studies, cold trap testing, corrosion studies, calibration of instruments etc. [13].

3.2 Development of Superconducting Magnet Technology

The technology for the indigenous fabrication of large Low Temperature Superconductor (LTS)/High Temperature Superconductor (HTS) magnets suitable for fusion reactor has been initiated. Compact and economical superconducting magnets for fusion device can be met by developing novel technologies such as the development of high current conductor, neutron resistant insulation system, conductor winding, structural materials, Nano-ohm class conductor joints, current leads, sensors for quench detection, quench protection system for large volume superconducting magnets and suitable superconducting test facilities. These developments not only support India's energy security, but also offer the excellent opportunities for Indian industries to participate in building the necessary infrastructure and expertise. In a major step forward, IPR has fabricated a 232 mm long HTS solenoid magnet with inner and outer diameters of 200 mm and 260 mm respectively using Rare-Earth Barium Copper Oxide (REBCO) tapes [14]. This has produced a central magnetic field of 3.52 T at 4.2 K for over 360 s, maintaining excellent cryo-stability. The current and field waveforms at 4.2 K have been measured for the 3 T HTS magnet.

3.3 Plasma Diagnostics

Plasma diagnostics plays a pivotal role in the domain of fusion research owing to its importance in the monitoring of machine performance and the understanding of the underlying physics. Currently, IPR has attained proficiency in the diagnosis of plasma through the use of diagnostics involving X-rays spectroscopy (both in the hard x-ray and soft x-ray region), UV visible spectroscopy, laser scattering and interferometry, microwave interferometry and reflectometry, Electron Cyclotron Emission, bolometry, and imaging in the IR and visible wavelength ranges. These diagnostic developments were performed in the plasma parametric range of an about 1 keV maximum temperature and an average plasma density of $\sim 10^{19}$ m^3. IPR collaborations through ITER and other machines have also provided significant know-how on various diagnostics.

3.4 Development of Cryogenic System

IPR has the indigenously developed helium plant with tested cooling power of (a) 200 W refrigeration at 4.5 K; (b) 600 W refrigeration at 18 K; (c) 80 l/h. helium liquefaction rate. These have been developed with about 90% indigenous content and gained significant experience in operating the large scale cryogenic system along with some design and manufacturing activities in cryogenic components. This plant will be used for several R&D activities like cryo-pump, superconducting magnet development, apart from using it in tokamak experiments. Development of Helium Refrigerator and Liquifier plants is an indigenous effort to make reliable cryogenic technology fully Atmanirbhar.

3.5 Auxiliary Heating Systems

Radio Frequency Heating and Current Drives (RF H&CD) such as Electron Cyclotron Resonance Heating (ECRH), Ion Cyclotron Resonance Heating (ICRH), Lower Hybrid Current Drive (LHCD) and Neutral Beam Injection (NBI) are critical technologies for achieving and sustaining the extreme conditions required for fusion energy in devices like tokamaks and stellarators. The frequency of ECRH and ICRH systems depend upon the toroidal magnetic field of operation of tokamaks and scales linearly with the toroidal magnetic field. In the case of Aditya-U, the fundamental frequency of operation of ECRH is ~42 GHz, and ICRH fundamental frequency for hydrogen excitation ~24 MHz. The frequency of operation of LHCD is ~3 GHz.

4 Mega-Science Project

ITER is a collaborative project of seven partners nations (European Union, India, China, Japan, South Korea, Russia and USA) aimed at demonstrating the scientific and technological feasibility of magnetically confined controlled thermonuclear fusion reactions as a source of energy for the future. ITER-India is the nodal agency for Indian deliverables to the ITER mega-science project in France. The in-kind contributions to ITER are largely manufactured/assembled and procured in India and delivered to the site in France. These deliverables involve a variety of systems involving mechanical and electrical engineering, electronics and physics. Some are engineering-intensive systems, like the Cryostat, an approximately 30 m tall and 30 m wide steel vessel. Other systems, such as large Cryogenic systems, Diagnostics, Ion Cyclotron Radio frequency Source or Diagnostic Neutral beam systems are R&D-intensive deliveries. India's participation in ITER enables access to full knowledge generated from the Project, hence, it will help to accelerating our own domestic fusion research program. It also helps Indian industries to enhance the industrial capabilities of the country by exposure to critical technologies required for fusion reactor.

LIGO project is a collaboration between the LIGO Laboratory (USA) and Indian institutions, including RRCAT, IUCAA Pune, and IPR. As a part of this consortium of the multidisciplinary mega-science project, IPR is responsible for Vacuum and Controls & Data Acquisition System (CDS) Infrastructure of this Observatory. A prototype of 20 m long cylindrical vacuum vessel integrated with 1250 mm gate valve (LI-VISTA: LIGO India Vacuum Integrated System Test Assembly) has been setup IPR for evaluation of 80 K Cryopump efficacy in trapping water molecules moving along the length of assembly. This 1:1 scale prototype manufacturing helped LIGO-India project to assess manufacturing feasibility, establish the manufacturing process and associated quality assurance plan.

5 Industrial and Societal Applications of Plasmas

IPR has been developing Industrial/societal applications of plasmas, as well as spinoffs from Fusion Technologies. Several plasma systems have been developed, such as eco-friendly waste disposal, medical/health applications, Surface engineering, Textile processing, defence, space, agriculture, water, green hydrogen, etc. Many of these technologies have been patented and transferred to industry. Plasma pyrolysis based common biomedical waste disposal technology "Raudra" is a major contribution to Aatmanirbhar Bharat and Swatchh Bharat Missions. Improved performance of industrial components like gears, mechanical tools, etc. by various plasma based surface engineering techniques (plasma nitriding, carburizing, coating etc.) have been demonstrated, and is available for industrial use. Plasma Nitriding System installed at FCIPT is now operated in GOCO (Government Owned, Contractor Operated) mode.

IPR has also developed Plasma Sterilization of medical devices, non-thermal plasma for disinfection and decontamination, Plasma Activated Water as a disinfectant for medical applications and faster growth of crops. An AI-based software called DeepCXR, developed by IPR and trained using ICMR data, has recently been cleared by ICMR and is now being rolled out for nation-wide TB screening in association with MoH&FW. On the technology innovation front, the AIC Plasmatech–a Sect. 8 company under DAE, supported plasma and allied technology start-ups in plasma science, fusion technology, advanced materials, cryogenics, vacuum systems, AI, robotics, clean energy, and allied engineering domains [15].

6 Summary

The institute has significantly advanced fusion technology, basic plasma science, and plasma applications through its varied R&D endeavours. IPR has successfully developed and validated reactor-relevant technologies, including blanket, divertor, superconducting magnets, cryogenics, plasma diagnostics, remote handling systems, etc. The tokamak program has generated manpower trained in tokamak experiments and critical areas in fusion technologies. IPR has outlined the proposal to build a tokamak (SST-Bharat) that will serve as the bridge for integrating the knowledge gained from ITER participation into our domestic program. The proposed program will bridge the gap between the current tokamak program in India and that required to start building our own DEMO reactor. FCIPT actively collaborates with MSMEs, start-ups, and large industries to transfer customized plasma solutions and promoting outreach to various societal benefits and industries. These technologies are low-cost substitutes for costly imported systems. These developments have thus a significant contribution towards Atmanirbhar Bharat initiative. Ten start-ups have been incubated and seven technology transfer agreements executed with Indian companies/start-ups. Additionally, outreach initiatives extended to encompass a wide range of educational institutions, including teacher training programs.

Acknowledgments. The authors would like to acknowledge all Division and Section Heads for their contribution to this paper.

References

1. https://www.ipr.res.in/
2. https://www.iterindia.in/
3. https://www.ligo-india.in/
4. https://www.ipr.res.in/outreach/index.html
5. http://www.plasmaindia.com/
6. https://www.iter.org/
7. http://www.cppipr.res.in/
8. Kumar, R., et al.: Study and analysis of the design considerations for controlling vertical plasma position in ADITYA-U tokamak. Fusion Eng. Des. **210**, 114736 (2025)
9. Raju, D., et al.: Igniting India's fusion future. Curr. Sci. **128**(11), 101090 (2025)

10. Yadav, B.K., et al.: Overview of the experimental helium cooling (EHCL) system. Fusion Eng. Des. **196**, 114006 (2023)
11. Belsare, S., et al.: Enhancement and comprehensive testing of interlock protection systems of high heat flux test facility at IPR. Fusion Eng. Des. **206**, 114588 (2024)
12. Abhangi, M., et al.: Neutron emission characterization of IPR 14 MeV neutron generator. Fusion Eng. Des. **204**, 114522 (2024)
13. Patel, A., et al.: First operation of LLMHD loop with electromagnet for R&D MHD experiments. Fusion Eng. Des. **207**, 114614 (2024)
14. Prasad, U., et al.: Design, fabrication and testing of a D-shaped high temperature superconducting magnet. https://doi.org/10.48550/arXiv.2601.10295
15. https://plasmatech.co.in/

Mathematical Sciences

Algorithms, Learning, and Randomness

Piyush Srivastava(✉)

Tata Institute of Fundamental Research, Mumbai, Maharashtra 400005, India
piyush.srivastava@tifr.res.in

Abstract. We survey recent and ongoing work at DAE institutions addressing the foundational aspects of learning. These include work on sampling algorithms for high-dimensional distributions, algorithms for optimization in dynamically changing environments, and on fundamental statistical notions such as (over-)parameterization and stability of models. An underlying unifying theme in many of these works is the question of understanding and exploiting the properties of randomness and their effect on algorithm design. This note is based on a short talk presented at the 2006 DAE Conclave, held at TIFR Mumbai in January, 2026.

1 Introduction

In this short note, I briefly describe some recent threads of research on algorithms and learning across DAE institutions. The hope is that this survey can help foster further collaborations across various institutions, both on theoretical aspects, and also on applications of such ideas in technology and the natural sciences. The note concludes with a discussion of a broad direction for future work.

Almost by necessity, a short survey such as this has to fall short of being comprehensive even in terms of the questions and problems discussed, and even more so in terms of chronicling the work of all the researchers working on these problems. The choice of examples I discuss below is often a product of the threads of work I am personally most familiar with, and I must apologize in advance for all the important work that may thereby remain undiscussed in this survey.

I start with a short (and necessarily incomplete) historical overview.

2 A Brief Historical Overview

The history of modern electronic computing at DAE institutions is almost as old as the history of electronic computing in the country: the TIFR Automatic Calculator (TIFRAC) is widely considered an early national milestone. Study of artificial intelligence at DAE institutions also dates back at least to the late 1960s/early 1970s: S. Ramani of the erstwhile "Computer Group" described an intriguing question-answering system in a 1971 paper [17]. Beginning in the 80 s, there was also research at DAE institutions on closely related foundational topics such as stochastic control (see, e.g., [4]). On the applications side, beginning in

G. B. Mohanty and V. S. Prabhudesai (Eds.): DAEC 2026, SPPHY 368, pp. 219–224, 2026.
https://doi.org/10.1007/978-981-92-2948-2_29

the 1990s, DAE researchers were already applying neural networks to high energy physics problems [3,6]. Use and analysis of such artificial intelligence (AI) and machine learning (ML) based methods for scientific applications continues to be an active focus of research across the DAE system. It is difficult to be comprehensive in describing the wide variety of this work. A few recent examples include design of metamaterials [15], detection of planetary candidates in astronomy [11], as well as the use of AI/ML techniques at various stages of large scale data analyses at the National Centre for Radio Astronomy (NCRA). It is worth noting that work in this direction has not remained confined to applications of existing tools, and has also contributed to critically appraising the challenges arising in the scientific use of such tools (see, e.g. [12]) and to the development of principled methods for evaluating the output of these tools (see, e.g., [7]).

In this note, however, I will focus more on theoretical aspects.

3 A Few Recent Examples

Over the last three to four decades, there has been significant research activity at DAE institutions on questions relevant to learning, from the perspective of areas such as stochastic processes, control theory, optimization, partial differential equations and numerical analysis. I now discuss a few recent examples of such work. An underlying theme here is the important and diverse role randomness plays: sometimes as a resource, and sometimes as a nuisance that has to be tackled.

Sampling and Integrating The following general problem of integration arises in various contexts: one is given a function f on a domain $\mathcal{D}$ along with a (probability) measure μ on $\mathcal{D}$, and wants to compute, or at least approximate, $\int_{x\in\mathcal{D}} f(x)\mu(dx)$. A generic approach to this is Monte Carlo integration: sample n points $x_1, x_2, \ldots, x_n$ independently from $\mathcal{D}$ according to μ, and approximate the integral by the sample average estimator $\frac{1}{n}\sum_{i=1}^{n} f(x_i)$.

The first problem that arises in implementing this approach is to have a procedure for sampling from μ. Even in the relatively simple case when $\mathcal{D}$ is a convex body and μ is the uniform distribution on $\mathcal{D}$, sampling uniformly, or even approximately uniformly, from μ can be a non-trivial problem [9]. Several algorithms have been proposed and analyzed for this problem, with a full theoretical understanding of the performance of some of the practically used algorithms still remaining elusive. Recent work at DAE has contributed to this direction by giving a new analysis method for the coordinate "hit-and-run" algorithm [14].

The second problem arises when, as in many applications in modern physics, the function f takes negative or even complex values. In this situation, samples corresponding to large magnitude values of f may cancel each other, thereby leading to a large variance in comparison to the magnitude of the expected value of the sample average estimator. Manifestations of this *sign problem* arise in many different fields, and typically one has to carefully exploit the special structure of μ, $\mathcal{D}$ and f in each specific application. As an example of this, recent work at DAE indicates that the idea of randomized polynomial identity testing from computational number theory can be used to speed up certain computations that arise in the context of lattice quantum chromodynamics (QCD) [5].

Optimization in a changing environment Applications such as control of robots often require optimization in a situation where constraints and objectives are not all known in advance, are only revealed progressively, and can even change in response to the actions of the controller. An abstract formulation of this is of (constrained) online convex optimization. Here, at each time t, a "soft" convex constraint g_t and a convex cost function C_t are revealed. The controller can then move to a state $\boldsymbol{x}_t$. The goal is to, over a time horizon T, minimize both

- the total "violation" of the constraints g_t, as well as
- the total cost $\sum_{t=1}^{T} C_t(\boldsymbol{x}_t)$.

In this setting, it is important to specify a useful benchmark against which the performance of the controller is measured. A standard choice is to use as a benchmark the best possible performance of a hypothetical "clairvoyant" algorithm that has the advantage that it knows all constraints and costs in advance, but is constrained in that it has to choose a single response $\boldsymbol{x}$ that it cannot vary over time. The goal then becomes to minimize a quantity called the *"regret"*, which is the difference between the total costs incurred by the clairvoyant benchmark and the controller, while also keeping the total constraint violation small. Recent work by A. Sinha and R. Vaze of TIFR gave an algorithm obtaining the best possible regret guarantees for this problem, under mild regularity conditions [18].

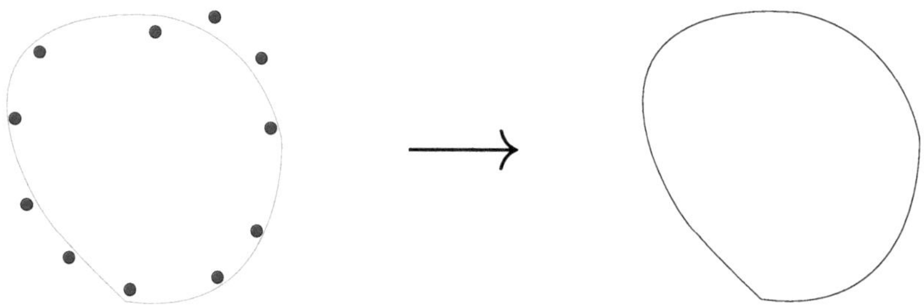

Fig. 1. Manifold learning under noise

Learning structure under large noise A useful abstraction in many applications is that the "ideal" data lies on a nice surface or manifold, but the observations are points from this ideal surface with random noise added to them. The goal then is to recover the ideal surface (with as much accuracy as possible) from such noisy observations (see Fig. 1). Such problems arise in contexts such as analysis of cryogenic electron microscopy (cryo-EM) data. Recent work involving H. Narayanan of TIFR has shown that under natural regularity conditions, manifolds can be learnt even in the setting when the noise in each ambient dimension is comparable to the diameter of the manifold [10].

Statistical foundations of learning Recent work by researchers at DAE has also considered the problem of fundamental statistical limits on machine learning models that may be expected to hold across a variety of models. Examples include studies of connections between over-parameterization and robustness [8] and of error rates in transformers for algorithmic tasks [16].

4 Potential Directions for the Future

State-of-the-art generative AI systems such as large language models (LLMs) are inherently stochastic in nature. At least at an intuitive level, it is therefore not surprising that by themselves, they can sometimes produce "hallucinations": outputs that look like well-structured text, but are flawed or simply wrong. Some techniques to reduce hallucinations take the form of attaching extra filters on top of the stochastic output of the LLM. Such filters can be of various types: sometimes they can just be in the form of another LLM that tries to "check" the output produced by the original LLM. Another common idea is the use of *tools*: for certain types of intermediate tasks that are likely to arise as an intermediate stage, the LLM may be given access to special purpose programs that it can call, rather than trying to perform the task itself (a rough analogy would be a human using a calculator to do arithmetic rather than doing mental arithmetic). An especially interesting case of this is the use of formal verification tools: the model is asked to generate its output in a formal language that is machine checkable, and errors found by the checker program can then in principle be fed back to the model, hopefully leading to a reduction in errors, or at the very least, to an early detection of hallucination. This is quite reminiscent of how humans perform mathematical reasoning: informal reasoning in natural language is interspersed with (partial) translation into formal symbolic notation and symbolic manipulation guided by formal rules. It is not surprising therefore that there has been much activity in adapting these methods to enhance the utility of AI models for pure mathematics [1]. One aspect of this activity has been to design models that can automatically translate existing mathematical proofs into a strict formal language (see, e.g., [13]).

An interesting question arising from this is to understand to what extent formalization-like methods can be used to enhance the utility of AI for natural sciences beyond pure mathematics. In other words, the question is to identify large classes of arguments or statements in physics, chemistry, and the life sciences that can be "automatically" verified. Unlike in the case of pure mathematics, this automatic verification may not be based purely on symbolic logic, and may depend upon a wider variety of "tools" such as database searches, first-principles numerical simulations, and even physical experiments (see, e.g., [2]). The problem then is to identify classes of such arguments and statements that are amenable to such checking, and are also useful for important downstream scientific tasks.

Acknowledgments. I acknowledge support of the Department of Atomic Energy, Government of India, under Project Nos. RTI4001 and RTI4014. I also thank Profs. Gagan Mohanty and Rajdeep Chatterjee for pointing me to the early work of Prof. T. Aziz.

References

1. Achim, T., Best, A., Bietti, A., Der, K., Fédérico, M., Gukov, S., Halpern-Leistner, D., Henningsgard, K., Kudryashov, Y., Meiburg, A., Michelsen, M., Patterson, R., Rodriguez, E., Scharff, L., Shanker, V., Sicca, V., Sowrirajan, H., Swope, A., Tamas, M., Tenev, V., Thomm, J., Williams, H., Wu, L.: Aristotle: IMO-Level Automated Theorem Proving (2025). https://doi.org/10.48550/arXiv.2510.01346
2. Anishchenko, I., Pellock, S.J., Chidyausiku, T.M., Ramelot, T.A., Ovchinnikov, S., Hao, J., Bafna, K., Norn, C., Kang, A., Bera, A.K., DiMaio, F., Carter, L., Chow, C.M., Montelione, G.T., Baker, D.: De novo protein design by deep network hallucination. Nature **600**(7889), 547–552 (2021). https://doi.org/10.1038/s41586-021-04184-w
3. Aziz, T.: Heavy flavor tagging from hadronic Z decays in the L3 detector using neural network technique. In: 4th International Workshop on Software Engineering and Artificial Intelligence for High-Energy and Nuclear Physics (1995)
4. Borkar, V.S.: A convex analytic approach to Markov decision processes. Probab. Theory Relat. Fields **78**(4), 583–602 (1988). https://doi.org/10.1007/BF00353877
5. Chakraborty, D., Srivastava, P., Kumar, A., Mathur, N.: Nuclear correlation functions using first-principle calculations of lattice quantum chromodynamics. Phys. Rev. D **110**(11), 114505 (2024). https://doi.org/10.1103/PhysRevD.110.114505
6. Chattopadhyay, S., Ahammed, Z., Viyogi, Y.P.: Application of neural network for photon-hadron discrimination in a preshower detector in high-energy heavy-ion collisions. Nucl. Instrum. Methods Phys. Res., Sect. A **421**(3), 558–566 (1999). https://doi.org/10.1016/S0168-9002(98)01213-3
7. Das, M., Venkatramani, R.: A mode evolution metric to extract reaction coordinates for biomolecular conformational transitions. J. Chem. Theory Comput. **20**(19), 8422–8436 (2024). https://doi.org/10.1021/acs.jctc.4c00744
8. Das, S., Batra, J., Srivastava, P.: A direct proof of a unified law of robustness for Bregman divergence losses. IEEE Trans. Inf. Theory **71**(8), 6340–6352 (2025). https://doi.org/10.1109/TIT.2025.3567076
9. Dyer, M., Frieze, A., Kannan, R.: A random polynomial-time algorithm for approximating the volume of convex bodies. J. ACM **38**(1), 1–17 (1991). https://doi.org/10.1145/102782.102783
10. Fefferman, C., Ivanov, S., Lassas, M., Narayanan, H.: Fitting a manifold of large reach to noisy data. J. Topol. Anal. **17**(02), 315–396 (2025). https://doi.org/10.1142/S1793525323500012
11. Gavankar, A., Mittal, T., Ninan, J.P., Hanasoge, S.: Machine learning for radial velocity analysis. I. Vision transformers as a robust alternative for detecting planetary candidates. Astron. J. **171**(1), 33 (2025). https://doi.org/10.3847/1538-3881/ae1115
12. Majila, K., Golatkar, O., Viswanath, S.: Integrative modeling with AlphaFold. Curr. Opin. Struct. Biol. **98**, 103239 (2026). https://doi.org/10.1016/j.sbi.2026.103239

13. Math, Inc.: Completing the Formal Proof of Higher-Dimensional Sphere Packing (2026). https://math.inc/sphere-packing
14. Narayanan, H., Rajaraman, A., Srivastava, P.: Sampling from convex sets with a cold start using multiscale decompositions. Probab. Theory Relat. Fields **191**, 1169–1232 (2025). https://doi.org/10.1007/s00440-024-01341-w, an extended abstract appeared in Proceedings of 55th Annual ACM Symposium on Theory of Computing (STOC) (2023)
15. Punjal, A., Garde, C., Prabhu, S.: Metamaterial design using distributed neural network (DiNN) approach. In: Proceedings of 47th International Conference on Infrared, Millimeter and Terahertz Waves (IRMMW-THz). pp. 1–2 (2022). https://doi.org/10.1109/IRMMW-THz50927.2022.9895992
16. Raju, S., Netrapalli, P.: A Model of Errors in Transformers (2026). https://doi.org/10.48550/arXiv.2601.14175
17. Ramani, S.: A language based problem-solver. In: Proceedings of the 2nd International Joint Conference on Artificial Intelligence (IJCAI), pp. 463–473. Morgan Kaufmann Publishers Inc., San Francisco, CA, USA (1971). https://dl.acm.org/doi/10.5555/1622876.1622924
18. Sinha, A., Vaze, R.: Optimal algorithms for online convex optimization with adversarial constraints. In: Advances in Neural Information Processing Systems (NeurIPS), vol. 37, pp. 41274–41302. Curran Associates, Inc. (2024)

Recent Theoretical Developments in Inverse Problems in India

Venkateswaran P. Krishnan(✉)

TIFR Centre for Applicable Mathematics, Bangalore, India
vkrishnan@tifrbng.res.in

Abstract. Inverse Problems research community in India is an up and coming group with varied interests ranging from those involved in theoretical investigations and development of numerical algorithms to those working at the interface of academia and industry. We give a broad overview of the research works of this relatively newer field in the Indian research landscape focusing on theoretical aspects.

1 Introduction

An inverse problem can be broadly defined as the question of determination of parameters modeling an object based on some measured or observed data from the exterior of the object. This requires probing the object by sending X-rays, pressure waves, sound waves, electric current etc., measuring the response exterior to the object, and analyzing it to determine the object parameters. Inverse problems and image reconstruction play important roles in several aspects of human life such as healthcare, border security, non-destructive testing, remote sensing and geophysical exploration to name a few.

The propagation of waves within an object is governed by partial differential equations (PDE) with the coefficients of the PDE representing some features of the object. Therefore, the mathematical formulation of an inverse problem involves the determination of the coefficients of a PDE from some measured data exterior to the object. Under further simplifying assumptions, an inverse problem can also be reformulated as the inversion of a linear or non-linear transform acting on functions or higher order objects such as vector fields, tensor fields etc.

The mathematical study of inverse and image reconstruction problems can have a huge impact on human life. Efficient mathematical algorithms or the study of a combination of several imaging modalities known as hybrid imaging can lead to better imaging devices in fields such as medicine and remote sensing. Artifact removal or reduction strategies and efficient algorithms to implement them can lead to fewer false positives in fields such as medical, seismic and radar imaging.

Traditionally, the strength of PDE research at several institutes and universities across India has been on the theoretical and numerical aspects of PDEs - the so-called study of the forward problem. That is, the study of well-posedness and

G. B. Mohanty and V. S. Prabhudesai (Eds.): DAEC 2026, SPPHY 368, pp. 225–235, 2026.
https://doi.org/10.1007/978-981-92-2948-2_30

qualititave behavior of solutions to PDEs and development of efficient numerical algorithms for them. In recent years, there has been significant development in the research of theoretical and numerical aspects of inverse problems in India.

We summarize a few of the recent contributions in this field from India. The article, naturally, reflects the author's interests and research contributions in this field and this write-up is not meant to be a comprehensive account of inverse problems research in India. The rest of the article is organized as follows. In Sect. 2, we will focus on Calderón inverse problem and several of its offshoots, and give the contributions of the Indian research community in this context. Our group has contributed greatly to research in integral geometry problems as well, and we describe this in Sect. 3. In Sect. 4, we give other approaches to the study of inverse problems in India. The chapter ends with a Conclusion in Sect. 5.

2 Calderón-Type Inverse Problems

In this section, we describe the original inverse problem proposed by Calderón in his seminal work [11]. This paper has been the starting point of several foundational works in this area of inverse problems. It has spawned several sub-problems, including a geometric version of Calderón problem and its relationship to cloaking, fractional Calderón inverse problem, Calderón-type inverse problems for higher order operators, inverse problems for hyperbolic and parabolic equations to name a few. Although significant strides have been made in all these areas, they still continue to be active areas of mathematical research, with several fundamental questions still open.

We begin with the discussion of the Calderón inverse problem. For a highly accessible introduction to this problem, we refer to the recent book on this topic; [19]. An object occupies a region Ω in $\mathbb{R}^n$ (in practice $n = 2, 3$) with boundary $\partial\Omega$. Let $\gamma(x)$ be the unknown conductivity of the object. An electric voltage is applied to the boundary and the steady state outgoing current flux is measured at the boundary. The goal is to recover $\gamma(x)$ (conductivity in the interior) from these measurements only on the boundary of the object. The solution to this problem has guided the construction of medical devices for breast cancer detection.

If $f(x)$ is the applied boundary voltage and $u(x)$ is the potential in the interior, then Ohm's law implies that

$$\nabla \cdot (\gamma(x)\nabla u(x)) = 0, \qquad x \in \Omega, u(x) = f(x), \qquad x \in \partial\Omega.$$

The measured data is the voltage to current map (or Dirichlet-to-Neumann map):

$$\Lambda_\gamma : f \mapsto \gamma \frac{\partial u}{\partial \nu}|_{\partial\Omega}.$$

If γ is known, then Λ_γ can be determined by the numerical computation of the solution of a PDE. We are interested in the inverse question: given Λ_γ (the data from the experiment), determine γ. Hence, our goal is the inversion of the (nonlinear) response map

$$\Lambda : \gamma \mapsto \Lambda_\gamma.$$

For an isotropic medium, γ and u are real valued functions, and there has been considerable progress in this case; see [11,59,78,80]. However, several challenging problems remain; see [37] and the references therein.

If the medium is anisotropic, the conductivity is modeled by a matrix function $\gamma(x) = (\gamma_{ij}(x))$ and the goal is still the inversion of the response map Λ_γ. In the anisotropic case, Λ_γ is not injective because a change of variables on Ω, keeping $\partial\Omega$ fixed, generates a new conductivity with the same response map as the original one. However, it is unknown whether this is the only obstruction to the injectivity of Λ_γ and how one could recover at least one of the $\gamma's$ in the set of conductivities with the same data set Λ_γ. The articles [17,38,48] and the references therein describe these and other open questions in this area. The non-injectivity of Λ_γ for the anisotropic problem provides a way to create metamaterials (see [34]) so that objects coated with this metamaterial are undetectable by electromagnetic means. We refer to [23–25,34,64] for the mathematical progress on cloaking and open problems in this area.

With this introduction to the Calderón problem, we next describe the contributions of mathematicians from India in the context of Calderón-type inverse problems. They have contributed substantially to the recently-emerged field of fractional Calderón problem, Calderón-type inverse problem for higher order operators with particular emphasis on the use of momentum ray transforms as a tool to study such problems, hyperbolic inverse problems both in overdetermined and determined settings, stability estimates for the Calderón problem to name a few.

2.1 The Fractional Calderón Problem

Since many physical and chemical processes are modeled by operators with fractional derivatives, fractional derivative based analogues of the Calderón inverse problem have received considerable attention in recent years. Let Ω be a bounded domain in $\mathbb{R}^n$, $n > 1$, and Ω_e be the exterior of Ω, that is $\Omega_e = \mathbb{R}^n \setminus \Omega$. For $0 < s < 1$, the fractional Schrödinger equation is

$$\left((-\Delta)^s + q\right) u = 0 \text{ in } \Omega, \quad u|_{\Omega_e} = f$$

and the fractional analogue of the Dirichlet-to-Neumann map is the map

$$\Lambda_q : f \to (-\Delta)^s \, u|_{\Omega_e}.$$

Similar to the original Calderón problem, one is interested in the unique determination of the potential q from Λ_q. The work [21] was one of the first in this direction.

Since then, this field has literally taken off and has produced a great number of excellent and highly technical recent results in inverse problems involving non-local Laplacian operators; for a partial list of works in this area, especially with contributions from Indian mathematicians on a few of these articles, see [7,16,18,20,35,65,69,72].

2.2 Calderón-Type Inverse Problems for Higher Order Operators

We consider the following higher order version of the Calderón problem. Let $\Omega \subset \mathbb{R}^n$ for $n \geq 3$ be a bounded domain with smooth boundary $\partial\Omega$. For $m \geq 2$, we consider the following polyharmonic operator $\mathcal{L}$ with lower order anisotropic tensor perturbations up to order m:

$$\mathcal{L}(x, D)u = \left((-\Delta)^m + \sum_{j=0}^{m-1} \sum_{i_1,\dots,i_{m-j}=1}^{n} A^{m-j}_{i_1\dots i_{m-j}}(x) D^{m-j}_{i_1\dots i_{m-j}} + q(x) \right) u = 0,$$

where $D^k_{i_1\dots i_k} = \frac{1}{\mathrm{i}^k} \frac{\partial^k}{\partial x_{i_1}\dots\partial x_{i_k}}$. The goal is to recover all the unknown perturbations (A^{m-j}) and the function q from certain boundary measurements, similar to the Dirichlet-to-Neumann map considered above. In the context of higher order version of the Calderón inverse problem, the first works were in [46,47]. The contribution from the Indian mathematics community has been in using momentum ray transforms [32,33,39,40] as a versatile tool to study such uniqueness questions; see [8,9,70]. Momentum ray transforms have found applicability in both linear and non-linear inverse problems involving higher order operators.

2.3 Hyperbolic and Parabolic Inverse Problems

We are interested in studying lower order perturbations of the wave/heat equation: Let Ω be a bounded domain in $\mathbb{R}^n$ with $n \geq 3$. For $T \gg 1$, consider the solid cylinder $Q = [0, T] \times \Omega$. In Q consider the following PDEs:

$$\left((\partial_t + A_0(t, x))^2 - \sum_{j=1}^{n} (\partial_j + A_j(t, x))^2 + q(t, x) \right) u(t, x) = 0 \text{ (hyperbolic)},$$

$$\left(\partial_t - \sum_{j=1}^{n} (\partial_j + A_j(t, x))^2 + q(t, x) \right) u(t, x) = 0 \text{ (parabolic)}.$$

where the coefficients $A = (A_0, A_1, \dots, A_n)$ and q are unknown and we are interested in the recovery of these coefficients from certain boundary measurements. One can consider a non-linear version of these PDEs as well, where the coefficients involved are non-linear. In the overdetermined case, that is, when the measured boundary data depend on more variables than the coefficients, there have been quite a few studies from the Indian mathematics community; see for instance [45,51,61,71,73].

In the setting of formally determined inverse problems for hyperbolic PDEs (the measured data depends on the same number of variables as the unknown coefficients in the hyperbolic PDE), in recent times there has been two works; see [42,43]. Formally determined inverse problems are harder to solve compared to the over-determined ones due to the fact that one has more boundary information than what is required to solve the inverse problem. The contents of both these works have been in deriving stability estimates for the recovery of the coefficients.

2.4 Stability Estimates in Calderón-Type Inverse Problems

Once uniqueness result for an inverse problem is established, one is naturally led to the question of how stable is the recovery of these coefficients from boundary information. This is important in practice since measurements have imprecision in the form of noise, machine round-off errors etc. The Calderón inverse problem is a highly ill-posed inverse problem; see [5,27,50]. In the context of several inverse problems of Calderón type, there have been a substantial amount of work on deriving stability estimates for the recovery of coefficients; see [12–15].

3 Integral Geometry Problems

As already mentioned, the propagation of waves within an object is governed by PDEs. However, under certain simplifying assumptions, the inverse problem can be recast as the inversion of integral transforms. There has been substantial success in the study of such integral geometry problems in the Indian inverse problems community and we give a quick review of some of the works.

Suppose f is a compactly supported smooth function on $\mathbb{R}^n$, x a point on $\mathbb{R}^n$ and ξ a unit vector in $\mathbb{R}^n$. Then (x, ξ) determines a unique line starting at x and traveling in the direction of the unit vector ξ, and we define the ray transform

$$(If)(x, \xi) = \int_{\mathbb{R}} f(x + t\xi)\, \mathrm{d}t.$$

In applications such as X-ray imaging or non-destructive testing, it has been useful to study the question of recovering f, given the ray transform at points (x, ξ) on some subset of $\mathbb{R}^n \times S^{n-1}$.

Suppose f is a compactly supported smooth function on $\mathbb{R}^n$. A point x in $\mathbb{R}^n$ and a unit vector ξ in $\mathbb{R}^n$ determine a unique line through x parallel to the unit vector ξ. The X-ray transform of f is the integral of f along these lines, that is

$$(If)(x, \xi) = \int_{\mathbb{R}} f(x + t\xi)\, \mathrm{d}t, \qquad x \in \mathbb{R}^n,\ \xi \in \mathbb{R}^n,\ |\xi| = 1.$$

The inversion of this transform, that is the recovery of f when $(If)(x, \xi)$ is known only for (x, ξ) in special subsets of $\mathbb{R}^n \times \mathbb{R}^n$, is crucial in X-ray imaging and non-destructive testing. However, the path taken by the beams may not be straight lines if the medium is not uniform and it has become quite important to study problems for non-uniform mediums, which we describe next.

When the medium is not uniform, the particles travel at different speeds at different points in the medium and the paths taken by the particles are not lines but geodesics (shortest travel time curves) for the medium. Suppose the medium is modeled as a Riemannian manifold M with a Riemannian metric g and f is a smooth function on M. For each $x \in M$ and each unit tangent vector ξ at x, let $\gamma_{x,\xi}$ be the unique geodesic through x with tangent ξ at x. The geodesic ray transform of f is the integral of f on the geodesics of M, that is

$$(If)(x,\xi) = \int_{\mathbb{R}} f(\gamma_{x,\xi}(t))\,\mathrm{d}t, \qquad x \in M,\ \xi \in T_x(M),\ g(\xi,\xi) = 1.$$

As before, the goal is the recovery of f, given $(If)(x,\xi)$ for all (x,ξ) in some special subset of TM. Such questions are important if f is a higher order object such as a vector field or a tensor field instead of a function. This is the field of tensor tomography. Tensor tomography comes up in earthquake prediction, photo-elasticity, polarization tomography and also when one tries to recover the metric of a Riemannian manifold with boundary from the lengths of the geodesics between its boundary points. The articles [30,63,74,77] describe the recent progress, challenges, and applications of this field. Two excellent introductions to the fields of tensor tomography are [74,76].

One can consider other types of integral geometry problems; where the sets of integration are hypersurfaces such as spheres. These have applications in fields such as photoacoustic, thermoacoustic and ultrasound tomography. There have been substantial contributions to these areas from the Indian mathematics community in the context of inverse problems in integral geometry. The integral geometry problems have been studied from an exact reconstruction as well as from an edge-detection or singularity reconstruction point of view. See [1,3,4,26,28,31,41,44,52,52,53].

4 Numerical, Bayesian and Other Approaches in Inverse Problems

There has been a strong community of mathematicians, physicists and engineers in India focused on numerical aspects of inverse problems, probabilistic and statistical approaches to study inverse problems and inclusion detection methods. We refer to [49,54,55,66,81,82] for works related to statistical and data assimilation approaches, to [60,75,79] for compressed sensing approaches in image reconstruction, to [2,6,22,36,62,67,68,83] for numerical analysis approaches to the study of inverse and image reconstruction problems, to [56–58] for regularization approaches in inverse problems, and [10,29] for inclusion detection methods in inverse problems.

5 Conclusion

Inverse problems have a wide variety of applications in multiple aspects of human life, including, healthcare, border security, remote sensing, material testing to name a few. The research community in India focusing on several aspects of these problems has grown substantially in the last few years. With this growth comes the need for this community to engage in constant dialogue for important cross-fertilization of ideas. Toward this goal, a yearly meeting of inverse problems researchers has been happening for the last two years with the third iteration of it set to happen at IIT Bombay in September of this year. These meetings have focused mainly on the theoretical study of inverse problems so far. However, as we have seen, this field has immense potential to have applicability in several applied aspects leading to the overall improvement of quality of human life in our country. For this reason, there is a critical and immediate need to have annual conferences in this field on a larger scale with wide participation from academia, government research labs, industry, practitioners and those from the machine learning community as well. For this to be a reality, the inverse problems community requires continued support in the form of funding for regular conferences and for academia-research labs-industry collaborations from policy makers.

Acknowledgments. The author thanks Prof. Gagan Mohanty, Prof. Vaibhav Pradhudesai, DAE and TIFR Mumbai for the kind invitation to give a talk at the IInd DAE Conclave held at TIFR Mumbai from January 14–18, 2026. The talk at this conclave formed the basis for this book chapter and he extends his sincere thanks for the encouragement to write this chapter. He thanks Prof. Sreekar Vadlamani for his constructive feedback on an earlier version of the article.

References

1. Abhishek, A., Mishra, R.K., Thakkar, C.: Generalized radon transforms over symmetric m-tensor fields in $\mathbb{R}^n$. Anal. Math. Phys. **16**(1), 26 (2026)
2. Adesokan, B.J., Knudsen, K., Krishnan, V.P., Roy, S.: A fully non-linear optimization approach to acousto-electric tomography. Inverse Probl. **34**(10), 104004 (2018)
3. Agrawal, D., Krishnan, V.P., Sahoo, S.K.: Unique continuation results for certain generalized ray transforms of symmetric tensor fields. J. Geom. Anal. **32**(10), 27 (2022)
4. Agrawal, D., Singhal, N.: d-plane transform: unique and non-unique continuation. Proc. Am. Math. Soc. **153**(9), 3841–3853 (2025)
5. Alessandrini, G.: Stable determination of conductivity by boundary measurements. Appl. Anal. **27**(1–3), 153–172 (1988)
6. Bhatt, R., Naik, N., Subramanian, V.K.: SSIM compliant modeling framework with denoising and deblurring applications. IEEE Trans. Image Process. **30**, 2611–2626 (2021)
7. Bhattacharyya, S., Ghosh, T., Uhlmann, G.: Inverse problems for the fractional-Laplacian with lower order non-local perturbations. Trans. Am. Math. Soc. **374**(5), 3053–3075 (2021)
8. Bhattacharyya, S., Krishnan, V.P., Sahoo, S.K.: Momentum ray transforms and a partial data inverse problem for a polyharmonic operator. SIAM J. Math. Anal. **55**(4), 4000–4038 (2023)
9. Bhattacharyya, S., Krupchyk, K., Kumar-Sahoo, S., Uhlmann, G.: Inverse problems for third-order nonlinear perturbations of biharmonic operators (2023)
10. Brander, T., Harrach, B., Kar, M., Salo, M.: Monotonicity and enclosure methods for the p-Laplace equation. SIAM J. Appl. Math. **78**(2), 742–758 (2018)

11. Calderón, A.P.: On an inverse boundary value problem. Comput. Appl. Math. **25**(2–3), 133–138 (2006)
12. Anupam Pal Choudhury and Horst Heck: Stability of the inverse boundary value problem for the biharmonic operator: logarithmic estimates. J. Inverse Ill-Posed Probl. **25**(2), 251–263 (2017)
13. Choudhury, A.P., Krishnan, V.P.: Stability estimates for the inverse boundary value problem for the biharmonic operator with bounded potentials. J. Math. Anal. Appl. **431**(1), 300–316 (2015)
14. Choudhury, A.P., Kumar, A.T.: High-frequency stability estimates for the linearized inverse boundary value problem for the biharmonic operator with attenuation on some bounded domains. Math. Methods Appl. Sci. **48**(1), 839–851 (2025)
15. Choudhury, A.P., Kumar, A.T.: High-frequency stability estimates for the inverse boundary value problems for the Schrödinger and the biharmonic operator with constant attenuation on certain bounded domains. J. Math. Anal. Appl. **556**(1), 130094 (2026)
16. Covi, G., Ghosh, T., Rüland, A., Uhlmann, G.: A reduction of the fractional Calderón problem to the local Calderón problem by means of the Caffarelli-Silvestre extension. J. Math. Pures Appl. **9**(35), 103886 (2026)
17. Dos Santos, D., Ferreira, C.E., Kenig, M.S., Uhlmann, G.: Limiting Carleman weights and anisotropic inverse problems. Invent. Math. **178**(1), 119–171 (2009)
18. Feizmohammadi, A., Ghosh, T., Krupchyk, K., Uhlmann, G.: Fractional anisotropic Calderón problem on closed Riemannian manifolds. J. Differ. Geom. **131**(2), 401–414 (2025)
19. Feldman, J., Salo, M., Uhlmann, G.: The Calderón Problem—An Introduction of Graduate Studies in Mathematics, vol 253. American Mathematical Society, Providence (2025)
20. Ghosh, T., Rüland, A., Salo, M., Uhlmann, G.: Uniqueness and reconstruction for the fractional Calderón problem with a single measurement. J. Funct. Anal. **279**(1), 108505 (2020)
21. Ghosh, T., Salo, M., Uhlmann, G.: The Calderón problem for the fractional Schrödinger equation. Anal. PDE **13**(2), 455–475 (2020)
22. Gottam, O., Naik, N., Pandey, P.K., Gambhir, S.: RBF level-set based fully-nonlinear fluorescence photoacoustic pharmacokinetic tomography. Inverse Probl. Sci. Eng. **29**(13), 3227–3260 (2021)
23. Greenleaf, A., Kurylev, Y., Lassas, M., Uhlmann, G.: Invisibility and inverse problems. Bull. Am. Math. Soc. **46**(1), 55–97 (2009)
24. Greenleaf, A., Kurylev, Y., Lassas, M., Uhlmann, G.: Approximate quantum and acoustic cloaking. J. Spectr. Theory **1**(1), 27–80 (2011)
25. Greenleaf, A., Lassas, M., Uhlmann, G.: On nonuniqueness for Calderón's inverse problem. Math. Res. Lett. **10**(5–6), 685–693 (2003)
26. Han, B., Paul, S., Shukla, N.K.: Microlocal analysis and characterization of Sobolev wavefront sets using shearlets. Constr. Approx. **55**(2), 661–704 (2022)
27. Heck, H., Wang, J.-N.: Optimal stability estimate of the inverse boundary value problem by partial measurements. Rend. Istit. Mat. Univ. Trieste **48**, 369–383 (2016)
28. Holman, S., Krishnan, V.P.: Ray transform of symmetric tensor fields on Riemannian manifolds with conjugate points. J. Geom. Anal. **35**(10), 329 (2025)
29. Hwang, G., Kar, M.: Reconstructing unknown inclusions for the biharmonic equation. J. Math. Anal. Appl. **530**(2), 127745 (2024)

30. Ilmavirta, J., Monard, F.: Integral geometry on manifolds with boundary and applications. In: The Radon Transform—The first 100 Years and Beyond of Radon Series Computer Applied Mathematics, vol. 22, pp. 43–113. Walter de Gruyter, Berlin (2019)
31. Jathar, S.R., Kar, M., Krishnan, V.P., Pattar, R.R.: Reconstruction, unique continuation and stability, weighted divergent beam ray transform (2024)
32. Shubham, R., Jathar, M., Kar, S., Venkateswaran, P., Krishnan, V.A., Sharafutdinov, S.: Normal operators for momentum ray transforms, I: the inversion formula. J. Four. Anal. Appl. **30**(5), 39 (2024)
33. Shubham, R., Jathar, M., Venkateswaran, P., Krishnan, V.A., Sharafutdinov, K.: Normal operators for momentum ray transforms, II: Saint venant operator (2024)
34. Jenkins, A.: Metamaterials: lost in space. Nat. Photon. **2**(11), 264 (2008). https://doi.org/10.1038/nphoton.2007.264
35. Kar, M., Lin, Y.-H., Zimmermann, P.: Determining coefficients for a fractional p-Laplace equation from exterior measurements. J. Differ. Equ. **406**, 338–365 (2024)
36. Kaushik, A., Yalavarthy, P.K., Saha, R.K.: Convergent Born series improves the accuracy of numerical solution of time-independent photoacoustic wave equation. J. Mod. Opt. **67**(9), 849–855 (2020)
37. Kenig, C., Salo, M.: Recent progress in the Calderón problem with partial data. In: Inverse Problems and Applications of Contemporary Mathematics, vol. 615, pp. 193–222. American Mathematical Society, Providence, RI (2014)
38. Kenig, C.E., Salo, M., Uhlmann, G.: Inverse problems for the anisotropic Maxwell equations. Duke Math. J. **157**(2), 369–419 (2011)
39. Venkateswaran, P., Krishnan, R., Manna, S., Kumar-Sahoo, V.A., Sharafutdinov, K.: Momentum ray transforms. Inverse Probl. Imag. **13**(3), 679–701 (2019)
40. Krishnan, V.P., Manna, R., Sahoo, S.K., Sharafutdinov, V.A.: Momentum ray transforms, II: range characterization in the Schwartz space. Inverse Probl. **36**(4), 045009 (2020)
41. Krishnan, V.P., Mishra, R.K.: Microlocal analysis of a restricted ray transform on symmetric m-tensor fields in $\mathbb{R}^n$. SIAM J. Math. Anal. **50**(6), 6230–6254 (2018)
42. Venkateswaran, P., Krishnan, R., Senapati, S.: Point sources and stability for an inverse problem for a hyperbolic PDE with space and time dependent coefficients. J. Differ. Equ. **342**, 622–665 (2023)
43. Krishnan, V.P., Rakesh, R., Senapati, S.: Stability for a formally determined inverse problem for a hyperbolic PDE with space and time dependent coefficients. SIAM J. Math. Anal. **53**(6), 6822–6846 (2021)
44. Krishnan, V.P., Stefanov, P.: A support theorem for the geodesic ray transform of symmetric tensor fields. Inverse Probl. Imag. **3**(3), 453–464 (2009)
45. Krishnan, V.P., Vashisth, M.: An inverse problem for the relativistic Schrödinger equation with partial boundary data. Appl. Anal. **99**(11), 1889–1909 (2020)
46. Krupchyk, K., Lassas, M., Uhlmann, G.: Determining a first order perturbation of the biharmonic operator by partial boundary measurements. J. Funct. Anal. **262**(4), 1781–1801 (2012)
47. Krupchyk, K., Lassas, M., Uhlmann, G.: Inverse boundary value problems for the perturbed polyharmonic operator. Trans. Am. Math. Soc. **366**(1), 95–112 (2014)
48. Liimatainen, T., Salo, M.: Nowhere conformally homogeneous manifolds and limiting Carleman weights. Inverse Probl. Imag. **6**(3), 523–530 (2012)
49. Madhuresh: Some Results on Contraction Rates for Bayesian Inverse Problems (2017)
50. Mandache, N.: Exponential instability in an inverse problem for the Schrödinger equation. Inverse Prob. **17**(5), 1435–1444 (2001)

51. Mishra, R.K., Purohit, A., Vashisth, M.: Inverse problem for a time-dependent convection-diffusion equation in admissible geometries. Res. Math. Sci. **12**(4), 30 (2025)
52. Rohit Kumar Mishra and Suman Kumar Sahoo: Injectivity and range description of integral moment transforms over m-tensor fields in $\mathbb{R}^n$. SIAM J. Math. Anal. **53**(1), 253–278 (2021)
53. Rohit Kumar Mishra and Suman Kumar Sahoo: The generalized Saint Venant operator and integral moment transforms. Proc. Am. Math. Soc. **151**(1), 189–199 (2023)
54. Mitra, A., Apte, A., Govindarajan, R., Vasan, V., Vadlamani, S.: A discrete view of the Indian monsoon to identify spatial patterns of rainfall. Dyn. Stat. Clim. Syst. **3**(1), dzy009 (2018)
55. Mitra, A., Apte, A., Govindarajan, R., Vasan, V., Vadlamani, S.: Spatio-temporal patterns of daily Indian summer monsoon rainfall. Dyn. Stat. Clim. Syst. **3**(1), dzy010 (2018)
56. Mondal, S.: On source identification problem for a parabolic equation with memory term. Appl. Anal. **105**(2), 354–397 (2026)
57. Mondal, S., Nair, M.T.: A linear regularization method for a parameter identification problem in heat equation. J. Inverse Ill-Posed Probl. **28**(2), 251–273 (2020)
58. Mondal, S., Nair, M.T.: A source identification problem in a bi-parabolic equation: convergence rates and some optimal results. Numer. Funct. Anal. Optim. **45**(3), 189–215 (2024)
59. Nachman, A.I.: Global uniqueness for a two-dimensional inverse boundary value problem. Ann. Math. II **143**(1), 71–96 (1996)
60. Naidu, R.R., Jampana, P., Sastry, C.S.: Deterministic compressed sensing matrices: construction via Euler squares and applications. IEEE Trans. Sig. Process. **64**(14), 3566–3575 (2016)
61. Nakamura, G., Vashisth, M., Watanabe, M.: Inverse initial boundary value problem for a non-linear hyperbolic partial differential equation. Inverse Probl. **37**(1), 015012 (2021)
62. Nanda, P., Reddy, G.M.M., Vynnycky, M.: Inverse two-phase nonlinear Stefan and Cauchy-Stefan problems: a phase-wise approach. Comput. Math. Appl. **123**, 216–226 (2022)
63. Paternain, G.P., Salo, M., Uhlmann, G.: Tensor tomography: progress and challenges. Chin. Ann. Math. Ser. B **35**(3), 399–428 (2014)
64. Pendry, J., Schurig, D., Smith, D.: Controlling electromagnetic fields. Science **312**, 1780–1782 (2006)
65. Quan, H., Uhlmann, G.: The Calderón problem for the fractional Dirac operator. Math. Res. Lett. **31**(1), 279–302 (2024)
66. Reddy, A.S., Apte, A., Vadlamani, S.: Asymptotic properties of linear filter for deterministic processes. Syst. Control Lett. **139**, 104676 (2020)
67. Reddy, G.M.M., Nanda, P., Vynnycky, M.: A decompositional approach for two-dimensional, two-phase, nonlinear inverse Stefan problems using the method of fundamental solutions. Stud. Appl. Math. **156**(2), e70184 (2026)
68. Roy, S., Venkateswaran, P., Krishnan, P., Chandrashekar, A.S., Murthy, V.: An efficient numerical algorithm for the inversion of an integral transform arising in ultrasound imaging. J. Math. Imag. Vis. **53**(1), 78–91 (2015)
69. Rüland, A., Salo, M.: Exponential instability in the fractional Calderón problem. Inverse Prob. **34**(4), 045003 (2018)
70. Sahoo, S.K., Salo, M.: The linearized Calderón problem for polyharmonic operators. J. Differ. Equ. **360**, 407–451 (2023)

71. Sahoo, S.K., Vashisth, M.: A partial data inverse problem for the convection-diffusion equation. Inverse Probl. Imag. **14**(1), 53–75 (2020)
72. Salo, M.: The Fractional Calderón Problem. https://arxiv.org/abs/1711.06103
73. Senapati, S., Vashisth, M.: Stability estimate for a partial data inverse problem for the convection-diffusion equation. Evol. Equ. Control Theory **11**(5), 1681–1699 (2022)
74. Sharafutdinov, V.A.: Integral geometry of tensor fields. Inverse and Ill-posed Problems Series. VSP, Utrecht (1994)
75. Munnu Sonkar, K.Z., Najiya, C.S., Sastry, S.: Interior reconstruction in tomography via prior support constrained compressed sensing. J. Inverse Ill-Posed Probl. **31**(1), 77–90 (2023)
76. Stefanov, P., Uhlmann, G.: Microlocal Analysis and Integral Geometry. Soon to be Realsed Book from AMS
77. Stefanov, P., Uhlmann, G.: Boundary and lens rigidity, tensor tomography and analytic microlocal analysis. In: Algebraic Analysis of Differential Equations from Microlocal Analysis to Exponential Asymptotics, pp. 275–293. Springer, Tokyo (2008)
78. Sylvester, J., Uhlmann, G.: A global uniqueness theorem for an inverse boundary value problem. Ann. Math. **125**(1), 153–169 (1987)
79. Theeda, P., Kumar, P.U.P., Sastry, C.S., Jampana, P.V.: Reconstruction of sparse-view tomography via preconditioned Radon sensing matrix. J. Appl. Math. Comput. **59**(1–2), 285–303 (2019)
80. Uhlmann, G.: The Cauchy data and the scattering relation. In: Geometric Methods in Inverse Problems and PDE Control of IMA Volumized Mathematical Application, vol. 137, pp. 263–287. Springer, New York (2004)
81. Vasan, V., Manisha, D.: Auroux: ocean-depth measurement using shallow-water wave models. Stud. Appl. Math. **147**(4), 1481–1518 (2021)
82. Vasan, V., Oliveras, K., Henderson, D., Deconinck, B.: A method to recover water-wave profiles from pressure measurements. Wave Motion **75**, 25–35 (2017)
83. Warbal, P., Saha, R.K.: Performance comparison of commonly used photoacoustic tomography reconstruction algorithms under various blurring conditions. J. Mod. Opt. **69**(9), 487–501 (2022)

Recent Advances in String Theory: Some Comments

Arnab Kundu(✉)

Theory Division, Saha Institute of Nuclear Physics, A CI of Homi Bhabha National Institute, 1/AF, Bidhannagar, Kolkata 700064, India
arnab.kundu@saha.ac.in

Abstract. We briefly review some salient ideas in string theory that have led to a plethora of research activities in recent times. These ideas revolve around quantum aspects of black holes and their connections with quantum chaotic dynamics and features of quantum entanglement. This is a summary of a talk delivered at the DAE Conclave 2026, held at TIFR, Mumbai.

Understanding the laws of Nature, within a formal and mathematical framework that can subsequently be also explored to obtain novel physical predictions, is a primary goal of modern research in theoretical physics. The observed Universe spans over a huge range of length scales: 10^{-35}–10^{26} m. On one end of it sits the tiniest Planck scale, 10^{-35} m, where the underlying laws of physics are still an open issue; on the other end, the Universe at large is presumably determined by laws of gravity, however, the ingredients of the observed Universe are largely unknown. Among the last type, dark matter and dark energy are two outstanding challenges to modern physics.

While it is desirable to formulate a consistent framework in which such questions can be posed and addressed systematically—this in fact was Einstein's dream—the lack of detailed observational access provides us with a further challenge. Even though remarkable strides have already been and are being made by the recent gravitational interferometry efforts as well as particle colliders, the window to Nature's laws is still best carved through mathematical consistency and a formal abstract perspective in physics.

Historically, string theory emerged from such an approach which is based on the foundational principles of general relativity and quantum mechanics. The basic premise of string theory is that the underlying fundamental constituent degrees of freedom of the Universe, at the smallest scale, are strings of various types. The resulting quantum spectrum provides us with the observable particle data across a wide variety of energy scales and with a variety of symmetry properties. However, it is important to note that, till date, the standard model of particle physics has not been obtained from string theory. Decades of theoretical advances in string theory, instead, have produced convincing evidence that string theory can be viewed as a framework to study the nature of space-time, black holes, gravity and quantum systems in general. In this spirit, string theory is

G. B. Mohanty and V. S. Prabhudesai (Eds.): DAEC 2026, SPPHY 368, pp. 236–240, 2026.
https://doi.org/10.1007/978-981-92-2948-2_31

similar to the framework of quantum field theory (QFT), of which the standard model of particle physics is one example.

Let us consider the best understood and observed physics till date. On large scales, this includes recent observations of black hole mergers, constraints on dark matter and dark energy contents. On small scales, we now have a very precise understanding on the dynamics of the standard model. It is natural to begin at this point and explore, e.g. two classes of questions: (i) What is the quantum mechanics of black holes? (ii) How do we understand the strong coupling dynamics in particle physics? Interestingly, string theory has provided us with a large class of examples of Holography [1–3] which is a framework that naturally brings together these two questions together.

Black holes are classical objects that are obtained by solving Einstein's equations. These objects are unique since light, and therefore no classical matter, can escape from within. Hence, there is an explicit tension with unitarity principle in quantum mechanics that postulates that information is never lost. It turns out that, it is possible to impose the Heisenberg uncertainty relation: $\Delta x \Delta p \geq \frac{\hbar}{2}$ near a classical black hole space-time. Here, Δx and Δp are the uncertainties associated to the position and the momentum of a particle and $\hbar$ is the Planck's constant. Interestingly, this can be systematically implemented for a wide class of examples in string theory, and one obtains an intriguing formula that tells us that a black hole, quantum mechanically, is a thermodynamic object, with an entropy [4,5]: $S = \frac{k_B c^3}{\hbar} \frac{A}{4G}$, where k_B is the Boltzmann constant, c is the speed of light, and G is the Newton's constant. The quantity A depends on the specific black hole: it is given by the area of the event-horizon, which is the classical point of no return for black holes. There are two extremely novel aspects of this formula: First, all known constants of Nature are present. Secondly, the thermodynamic entropy is area-extensive. This feature essentially gives rise to the idea of Holography which sometimes takes a precise form of duality between a purely gravitational system in $(d+1)$-dimensions and a QFT-system in d dimensions.[1]

On one hand, the existence of a thermodynamic entropy demands a statistical mechanical description of black holes, which can be explicitly constructed within string theory for a certain class of black holes with a high degree of symmetry. Nonetheless, from a modern perspective of thermalization, it begs for a better understanding of the thermodynamic behaviour in terms of an underlying dynamics, e.g. quantum chaos, and the production of fine-grained entropies, e.g. Renyí entropies. Interestingly, ideas of quantum chaos, dynamics of fine-grained entropies have emerged as very powerful ways to think about a general quantum mechanical system with a large number of degrees of freedom. Recent years have subsequently seen a surge of research activities that are highly synergetic with other branches of theoretical physics.

We will now introduce the broad ideas and describe how they naturally fit in the stringy framework. Towards that, let us first recall that string theory provides

[1] Note that, the best understood examples of this duality are framed in the so-called anti de-Sitter (AdS) spacetime, which has a time-like boundary. However, how such a duality may be realized in other cases is an active area of current research.

us with a large class of black hole geometries, in which a special symmetry, known as the conformal symmetry, plays a crucial role. This symmetry is to be understood as a symmetry under a scaling transformation. These class of models are largely based on studying dynamics in an AdS-geometry, which in two-dimensions, reduce to the Poincaré disk, as shown in Fig. 1. All subsequent physical questions can now be framed in terms of quantum dynamics on this hyperbolic geometry, with a large number of degrees of freedom.

Let us begin with the idea of quantum chaos. One way to characterize quantum chaotic dynamics is, especially early time dynamics, through the out-of-time-order correlators (OTOCs). This emerges from the quantum analogue of sensitivity of classical trajectories to initial conditions: $\frac{\partial x(t)}{\partial x(0)} \implies \langle \hat{x}(t)\hat{x}(0)\hat{x}(t)\hat{x}(0)\rangle$ where $\hat{x}$ is the corresponding self-adjoint operator that acts on the Hilbert space. Of particular importance is the thermal state, in which case, such correlators, for a quantum chaotic dynamics, contain an exponential-in-time behaviour: $(1/N)e^{\lambda t}$, where N counts the number of degrees of freedom and λ denotes a Lyapunov exponent. Unitarity imposes a maximal bound for the Lyapunov exponent: $\lambda \leq \lambda_{\rm max}$, with $\lambda_{\rm max} = 2\pi k_B T/\hbar$ [7]. This bound is shown to be saturated for black holes. Related to this is the notion of how fast information can spread quantum mechanically, which is characterized by the so-called scrambling time. Black holes can be shown to be the fastest scramblers in Nature, for which the scrambling time $t_{\rm scr} \sim \log N$. All these results add up to the qualitative understanding of quantum black holes as the most thermal object

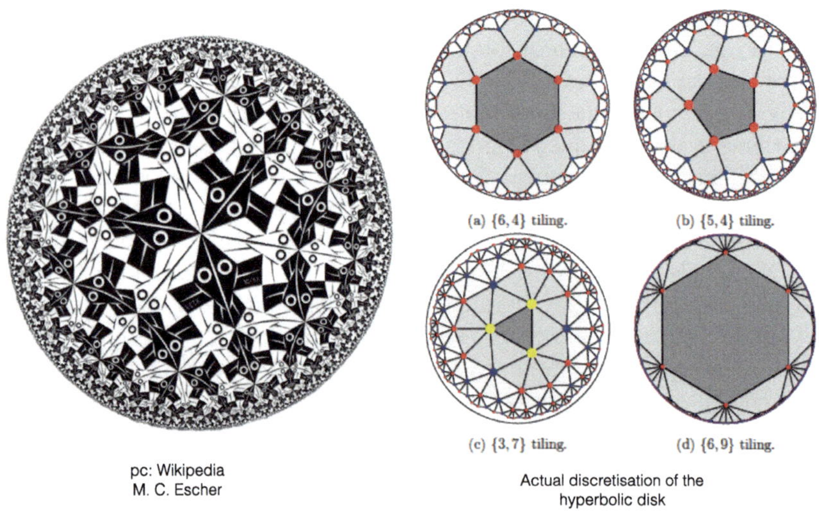

Fig. 1. On the left, we present an artist's rendition of the hyperbolic geometry. On the right, we exhibit explicit discretization of this geometry. Studying a quantum system on this discrete lattice, in the continuum limit, yields a QFT that captures semi-classical physics of black holes. The figure on the right is taken from [6].

in Nature, and dynamics on the hyperbolic disk provides us with a controlled framework for addressing the same.

There is a complementary view on the dynamics of thermalization: one in terms of fine-grained entropic quantities. This generally leads us to the notion of entanglement monotones, of which the von Neumann entropy is a popular measure. In a quantum mechanical system, with a factorizable Hilbert space $\mathcal{H}_{AB} = \mathcal{H}_A \otimes \mathcal{H}_B$, one considers a bi-partite density matrix ρ_{AB}. Upon a partial tracing, $\rho_A = \mathrm{tr}_B \rho_{AB}$, a reduced density matrix is obtained and subsequently the von Neumann entropy is defined by $S_{\mathrm{vN}} = -\rho_A \log \rho_A$. There is a precise duality, known as the Ryu-Takayanagi (RT) prescription [8], between geodesics on the hyperbolic disk between two boundary anchored points, which computes the von Neumann entropy of the dual Holographic QFT. This is illustrated in Fig. 2.

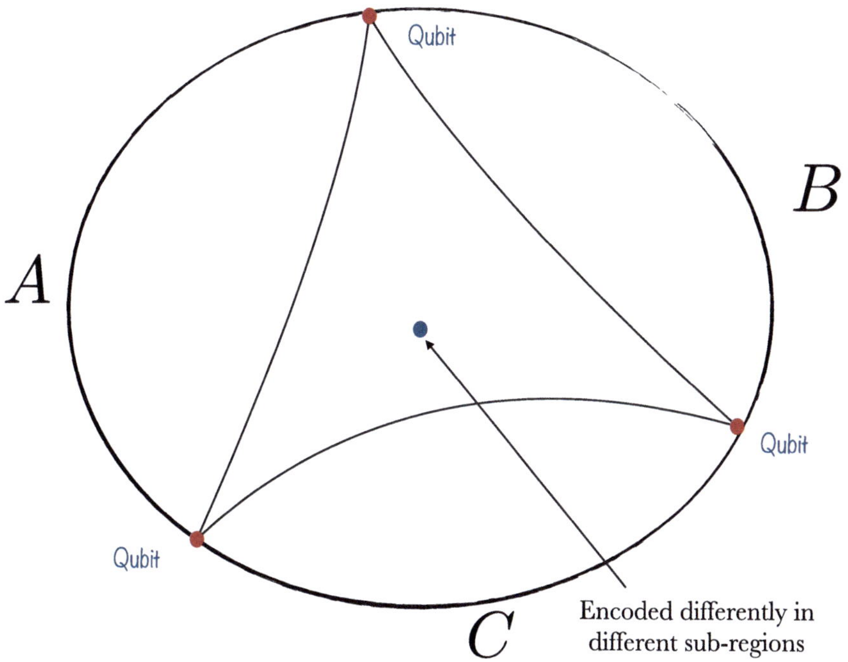

Fig. 2. The physical qubits, denoted by the red dots, are placed at the boundary of the hyperbolic disk while the logical qubit, denoted by the blue dot, is located in the bulk. Geodesics connecting the end points of either A, B or C sub-system does not contain the logical qubit in it, but a geodesic connecting the end points of e.g. $A \cup B$ does contain the logical qubit.

The RT-prescription translates a gravitational description into a QFT description. In particular, it hints towards the emergence of space-time through entanglement between separated points on the boundary. For example, in Fig.

2, the bulk point can be reconstructed by considering entanglement between the region $A \cup B/B \cup C/C \cup A$ and the rest, at the boundary. The structure of the hyperbolic disk ensures that the logical qubit, i.e. the bulk point, can be represented by different physics qubits, i.e. inequivalent regions at the boundary. Such a correspondence is deeply connected to the notion of error correcting codes in quantum information theory [9].

In conclusion, string theory has morphed into a structure that can no longer be termed as a "theory", but a framework where formal and abstract techniques are being used to address theoretical questions in physics, in general. For example, to better understand quantum aspects of the accelerating Universe, towards a potential rewriting of the Feynmann rules in QFT, understanding better the physics of phase transitions, potential signatures of quantum gravitational effects in gravity waves (e.g. gravitational memory), black holes and its connections to fluid dynamics and formal mathematical aspects in general.

References

1. Hooft, G.T.: Dimensional reduction in quantum gravity. Conf. Proc. C **12**, 930308 (1993)
2. Susskind, L.: The World as a hologram. J. Math. Phys. Ser. **36**, 6377 (1995). https://doi.org/10.1063/1.531249
3. Maldacena, J.M.: The Large N limit of superconformal field theories and supergravity. Adv. Theor. Math. Phys. Ser. **2**, 231 (1998). https://doi.org/10.4310/ATMP.1998.v2.n2.a1
4. Bekenstein, J.D.: Black holes and entropy. Phys. Rev. D Ser. **7**, 2333 (1973). https://doi.org/10.1103/PhysRevD.7.2333
5. Hawking, S.W.: Particle creation by black holes. Commun. Math. Phys. Ser. **43**, 199 (1975). https://doi.org/10.1007/BF02345020
6. Basteiro, P., Di Giulio, G., Erdmenger, J., Karl, J., Meyer, R., Xian, Z.-Y.: Towards explicit discrete holography: aperiodic spin chains from hyperbolic tilings. SciPost Phys. **13**, 103 (2022)
7. Maldacena, J., Shenker, S.H., Stanford, D.: A bound on chaos. JHEP **08**, 106 (2016)
8. Ryu, S., Takayanagi, T.: Holographic derivation of entanglement entropy from AdS/CFT. Phys. Rev. Lett. Ser. **96**, 181602 (2006). https://doi.org/10.1103/PhysRevLett.96.181602
9. Almheiri, A., Dong, X., Harlow, D.: Bulk locality and quantum error correction in AdS/CFT. JHEP Ser. **04**, 163 (2015). https://doi.org/10.1007/JHEP04(2015)163

Basic Science Education, Training, and Outreach

Promoting STEM Equity and Excellence: From National Leadership to Smartphone-Based Experimentation

Praveen Pathak(✉) and Arnab Bhattacharya

Homi Bhabha Centre for Science Education (HBCSE), Tata Institute of Fundamental Research, Mumbai, India

{praveen,arnab}@hbcse.tifr.res.in

Abstract. This paper (This paper is based on two talks delivered in the basic science education, training and outreach vertical at the 2nd DAE conclave held in TIFR, Mumbai from Jan. 14–18, 2026) presents an overview of the multi-faceted efforts at the Homi Bhabha Centre for Science Education (HBCSE) to promote equity and excellence in STEM education across India. We detail a wide gamut of activities, ranging from leadership of national talent nurture programs like the Science and Mathematical Olympiads to grassroots outreach initiatives. A significant development in recent times has focused on democratizing smartphone access to science by leveraging the "laboratory in your pocket", the now ubiquitous smartphone. By utilising internal sensors and low-cost materials, we demonstrate how advanced science concepts can be taught in under-resourced settings, bridging the gap between elite institutional facilities and the typical school classroom.

Keywords: STEM education · Science and Mathematics Olympiads · smartphones

1 Introduction

For over 50 years, HBCSE, a National Centre of the Tata Institute of Fundamental Research (TIFR), has been dedicated to improving science and mathematics education from primary school to the postgraduate level. HBCSE occupies a unique niche in the TIFR ecosystem with its core mandate based on the two pillars of equity and excellence, to encourage the growth of scientific literacy, nurture scientific talent, while also reaching under-served segments of the educational system.

HBCSE's activities [1] are diverse and impactful, as seen in the collage in Fig. 1. In particular the centre is responsible for

Talent Nurture & Olympiads For over 25 years, HBCSE has been the nodal centre of the Government of India for the national Olympiad programme in Science and Mathematics, identifying and training students who represent India in the International Olympiads. Indian teams consistently achieve stellar results across all subjects with an

G. B. Mohanty and V. S. Prabhudesai (Eds.): DAEC 2026, SPPHY 368, pp. 243–248, 2026.
https://doi.org/10.1007/978-981-92-2948-2_32

Fig. 1. A collage of various HBCSE activities ranging from international Science Olympiads to rural outreach, and from teacher professional development to educational game design.

enviable record in the medals tally. Figure 2 shows the improvement in the performance of the Indian team in the various Science and Mathematical Olympiads with time.

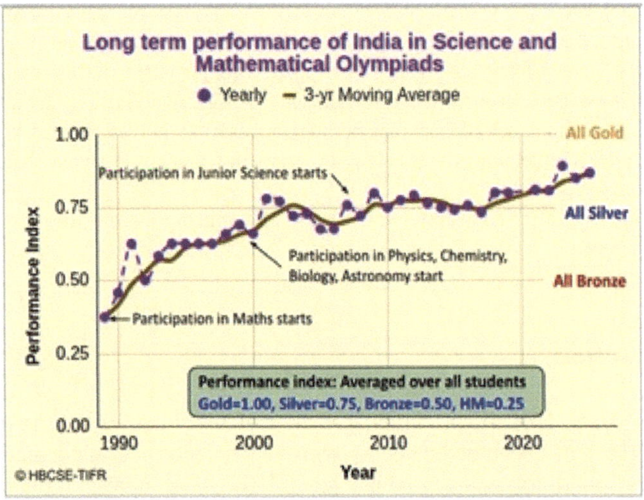

Fig. 2. Summary of India's performance in the international science and mathematical

Vigyan Pratibha This programme nurtures aptitude in science and mathematics among class 8–10 students, particularly those from disadvantaged groups. This programme is focused on teachers from the Kendriya Vidyalaya, Jawahar Navodaya Vidyalaya and Atomic Energy Central Schools, apart from Zilla Parishad and low-resource government schools. The programme is spearheaded by HBCSE with participation from regional centres across India allowing a national footprint. Several learning units have been developed and translated into regional languages.

National Initiative in Undergraduate Science (NIUS) This bespoke programme provides UG students in Physics, Astronomy, Biology, and Chemistry exposure to advanced research, hands-on experiments, and sustained mentorship by scientists across the country.

NCERT's New Textbooks Several members from HBCSE, including past members and alumni, are playing a key role in the development of the Science, Mathematics, Fun-Based Learning Activities, and other books in line with the National Education Policy (NEP 2020) and the National Curriculum Framework for School Education (NCFSE 2023)

Outreach and Innovation Initiatives like Vigyan Vidushi – specifically for women MSc students in physics – help address the gender gap in the sciences, while our Anandi newsletter reaches thousands of rural students across several districts of Maharashtra, and the Aamchi Prayogshala programme brings laboratory experiences to low-resource school settings in the MMR region. HBCSE's activities are unique since they build upon the experience of a rich and vibrant research base in science and mathematics education at the centre.

2 Smartphone-Based Experiments to Democratise Access to Science

A primary challenge in the Indian STEM education ecosystem is the **lack of functional laboratories.** National reports indicate that while many schools have "Integrated Science Labs," a very small percentage actually have working equipment enabling regular laboratory sessions or possess advanced facilities like "Tinkering Labs". To address this, we advocate for the use of smartphones as versatile scientific instruments. Driven by rapid rural adoption, India is projected to have approximately 1 billion smartphone users sometime in 2026, with over 75% of mobile subscribers using smartphones [2]. A modern smartphone contains a sophisticated suite of sensors, including:

Accelerometer and Gyroscope for mechanics.
Magnetometer for studying magnetic fields.
Light and Sound sensors for optics and acoustics.
Pressure and Temperature sensors.

By using free, open-source software like ***phyphox***—which HBCSE is helping to translate into Indian languages like Hindi—any student with a phone can turn their home into a high-precision lab. This approach ensures the authenticity of data and allows students to conduct experiments at minimal cost, often without requiring a fancy laboratory setup, and truly democratising access to science (Fig. 3).

3 Smartphone-Based Experiments at HBCSE

HBCSE has developed and published several innovative experimental setups that use smartphones to investigate complex physical phenomena. These are often integrated into National Olympiad examinations to provide opportunities for students to engage with real-world data. Key examples include:

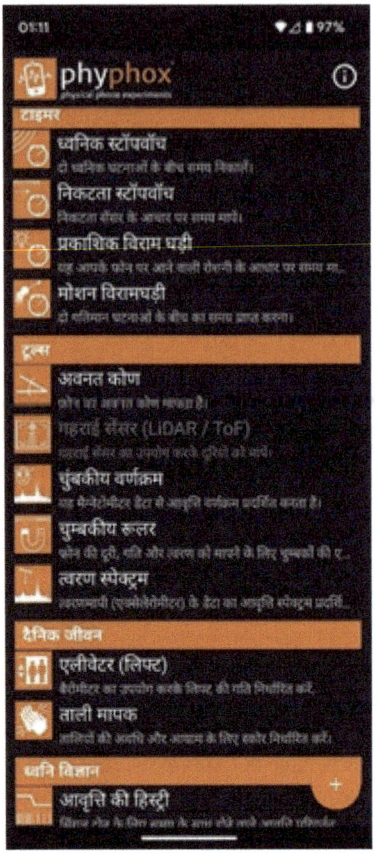

Fig. 3. The Hindi interface for PhyPhox, developed at HBCSE

Acoustics Students use the audio sensor to measure the frequency of sound as a tall container is filled with water. This reveals a rich spectrum of resonant frequencies that change as the container fills with water [3]. The experiment allows students to derive the speed of sound from the intercept of a $1/f$ vs. time plot for the different harmonics. Similarly, the smartphone also allows the visual delineation of nodes and antinodes inside tubes [4].

Mechanics The accelerometer can be used to analyze a range of motion, including cases of non-uniform motion, such as the acceleration and deceleration of an elevator, which can then be processed (using AI tools or manual calculation) to estimate velocity and displacement. The sensitivity of the accelerometer also allows it to be used to measure very small movements, such as the heartbeat of the human body.

Magnetism Magnetometers are used as precise timers to measure gravitational acceleration by tracking a falling magnet [5]. We also use the magnetometer to track the oscillations of magnetically coupled pendulums, providing a cost-effective way to visualize complex wave behaviour and dipole moments [6].

Optics A phone's high-resolution display can serve as an optical bench [7]. By placing a water drop on the screen (acting as a lens), students can image individual RGB pixels. Furthermore, using a laser on the display allows for the study of diffraction and reflection patterns since the pixel matrix serves as a two-dimensional diffraction grating [8].

To support these experiments, we have designed a low-cost kit consisting of everyday items like a CD, laser, fidget spinner, glass slab, and magnets, to allow the phone to be used for a range of experiments in a pedagogical environment.

4 Experiments Showcased at the Conclave

The authors were delighted to be able to showcase the expansive range of experiments enabled by the sensors in the smartphone to the delegates attending the DAE conclave in the Homi Bhabha Auditorium. Interested participants were requested to download the PhyPhox app on their smartphones and join in the experiments being demonstrated on stage. Starting with an introduction to the 3-axis accelerometer sensor, the smartphone was used to first measure 'g' the gravitational attraction due to the earth, and then creatively used to measure human reaction time by dropping and catching the phone. After an introduction to timed mode measurements, the sensitivity of the accelerometer was then exemplified by using it to measure heartbeat and breathing motions of the chest, and then quite dramatically measure microtremors of a supposedly "steady" hand. Another surprising experiment was the use of the phone's pressure sensor to measure atmospheric pressure on the floor, and on the top of a volunteer's head. The pressure sensor is sensitive enough to detect the minor variation (of barely 10–20 Pa) which can then be used to estimate the height of the individual! Experiments with the magnetometer were used to determine the local North-South direction. The frequency sensor, and the "audio scope" were demonstrated by having people whistle or sing, and the Doppler effect was shown by rotating the phone rapidly. These engaging and enjoyable experiments were enthusiastically appreciated by the audience (Fig. 4).

5 Conclusion

The work at HBCSE demonstrates that in science education, excellence and equity need not be mutually exclusive. By combining large-scale national nurture programs with innovative, low-cost technological solutions like smartphone-based science experiments, we show how the imagination of young minds can be ignited regardless of their socio-economic background. Our goal is to ensure that every student has the tools to satisfy their curiosity and develop a lifelong engagement with science.

Fig. 4. (a) the authors conduct an experiment on stage with the live data from the phone accelerometer being displayed on screen, (b) the audience enthusiastically participated in the session conducting experiments on their own smartphones

Acknowledgments. We acknowledge the use of an AI tool, NotebookLM, to help with English proofing of this manuscript. The work at HBCSE was supported by the Department of Atomic Energy, Government of India, through HBCSE- TIFR under Project Identification No. RTI4001.

References

1. www.hbcse.tifr.res.in
2. https://www.pib.gov.in/PressReleasePage.aspx?PRID=2132330&reg=3&lang=2
3. Maddur, M., Bhattacharya, A., Pathak, P.: Studying temperature-dependent sound speed in a constantly-filling beaker using a smartphone. Phys. Teach. **62**, 260–263 (2024)
4. Pathak, P., Maddur, M., Bhattacharya, A.: Investigating resonance in paper pipes and Rubens tube with a smartphone. Phys. Teach. **63**(7) (2025)
5. Pathak, P., Maddur, M.: Using smartphone magnetometers as timers to measure g. Phys. Teach. **63**(6), 499–502 (2025)
6. Pathak, P., Patel, Y., Raj, R.: Magnetically coupled oscillators and a smartphone. Phys. Teach. **62**, 234–238 (2024)
7. Freeland, J., et al.: Water droplet on a smartphone. Phys. Teach. **58**, 360 (2020)
8. Maddur, M., Shah, H., Pathak, P.: From pixels to patterns: decoding smartphone display properties through diffraction, reflection, and refraction. Am. J. Phys. **94**(1), 40–50 (2026)

Infusing Creativity and Collaboration in Teaching

Sonia Vikas Kapoor(✉)

Atomic Energy Central School No.2, Mumbai, India
soniakapoor2108@gmail.com

Abstract. Atomic Energy Education Society (AEES) is a dynamic educational system committed to excellence, inclusivity, and value-based education, promoting an inclusive ecosystem with equitable learning opportunities. Aligned with NEP 2020, it focuses on competency-based learning, flexibility and skill development. AEES encourages academic excellence and innovation through creative teaching-learning materials while supporting holistic development and experiential learning that connects classroom knowledge with real life. Students have earned recognition at national-level platforms, and many teachers have been honoured with the National Award for Teachers. Active alumni engagement and community participation further strengthen the learning environment, helping nurture responsible and capable learners.

Keywords: value-based education · teaching-learning materials · holistic development · and experiential learning

1 Introduction

Vision and Legacy Atomic Energy Education Society (AEES) has evolved into a robust system committed to academic excellence, inclusivity, and value-based education.

The journey of AEES began in 1969 in Anushakti Nagar, Mumbai with a single school. Today, it has transformed itself into a Pan-India educational network spanning 16 centres and administering approximately 30 schools and junior colleges. It nurtures over 27,000 students supported by a dedicated workforce of over 1,700 staff members, the majority of whom are teachers serving as the backbone of the system.

AEES operates with a clear vision: nurturing responsible, competent and value-oriented learners. "Let education help us see truth and virtue" is the guiding principle and it is inscribed on the emblem of AEES.

2 Inclusivity and Social Responsibility

AEES has an inclusive ecosystem. Admission is also extended to Non-DAE wards based on carefully designed criteria. RTE (Right to Education) admissions are also conducted in a transparent manner and strictly as per the norms.

G. B. Mohanty and V. S. Prabhudesai (Eds.): DAEC 2026, SPPHY 368, pp. 249–252, 2026.
https://doi.org/10.1007/978-981-92-2948-2_33

From among these RTE students, some are selected for the Special Educational Enrichment Program (SEEP) facility, where special support facilities are provided to the students on the basis of well-defined parameters.

3 Academic Excellence and Innovation

Excellence is the watchword for AEES, and the organization continuously raises the bar to meet changing educational needs.

Academic Performance There is a consistent upward trend in Class X and Class XII Board Examination results. This success is measured not just through pass percentages but also through the Quality Index (QI) and Excellence Index (EI), reflecting a sustained focus on academic rigor. Students pursue diverse career pathways, reflecting the system's readiness to support varied academic trajectories.

Integration of Technology and TLMs AEES utilizes the ASTER Studio, equipped with technical infrastructure for creating good quality e-content and learning resources. Teaching–Learning Materials (TLMs) are made by teachers using low-cost or scrap materials, ensuring they are portable, bilingual, and inclusive.

Primary Pitara An indigenous resource box at AECS-2, Mumbai has been made from scrap wood. It contains self-made TLMs and reading cards accessible to both teachers and students.

Open Library This library relies on student-donated books in Hindi, English, and Marathi, fostering a feeling of "joy of sharing" while celebrating Bhartiya Bhashas.

The book holders have been made using scrap wood in school.

4 Holistic Development and Experiential Learning

The curriculum at AEES seeks to shape individuals with clarity of thought and a strong sense of individuality. The focus has shifted from STEM to STEAM, integrating arts, language, and creativity alongside science.

4.1 Experiential Initiatives in some of the Atomic Energy Central Schools in the Recent Past-

- Nature Club: Students developed a self-sustaining nursery, growing saplings which are sometimes gifted as souvenirs in hand-painted pots.
- Students grew spinach in a school garden and shared a collective meal, instilling a respect for food and responsibility.
- Mud to Masterpiece: This activity involves inviting potters to teach students how to create eco-friendly clay *diyas*.
- Building as Learning Aid (BALA): Schools use the building itself as a resource, where walls are painted with educational concepts for informal, sight-based learning.
- Educational Visits: AEES supports visits of students to different institutions and places to support learning through real world exposure.

4.2 National Level Achievements

AEES students maintain a strong presence in national competitions:

- Rashtriya Bal Vaigyanik Pradarshini: 18 exhibits of AEES were presented in the recently conducted RBVP at Bhopal.
- Students also actively take part and get recognition in different competitions like Inspire Manak Award, CBSE Science Exhibition, IOQM, VVM, Green Olympiad, Aryabhata Ganit Challenge, M. R. Pai Memorial English Elocution, Painting Competition by Bureau of Energy Efficiency and much more.
- Sports and Co-curricular activities further add to this rich canvas.
- A wide range of competitions are conducted regularly through inter AEES Competitions.
- A good number of teachers have also been honoured with the National Award for Teachers, highlighting their dedication and excellence.

5 Teacher Empowerment and Institutional Growth

AEES recognizes that educators must grow alongside the system.

- Mandatory Training: Every educator undergoes 50 h of mandatory training annually (25 h online and 25 h offline) to stay aligned with NEP 2020 and emerging trends in education.
- Leadership and Research: Regular meetings and conferences are conducted for the staff members and parents to help encourage a shared vision. The first AEES Conclave showcased academic leadership through research papers, poster presentation, panel discussions and exhibition of teaching aids.
- Professional Contributions: Some AEES teachers serve as CBSE Resource Persons and train teachers of other CBSE schools.
- SHVR (Swachh Evam Harit Vidyalaya Rating) is an initiative of Ministry of Education. Atomic Energy Central Schools participate in it to improve the existing practices and also to demonstrate their commitment towards sanitation and hygiene.

6 Community Engagement and Collaborations

- Institutional Partnerships: AEES collaborates with the Tata Institute of Social Sciences (TISS) for internships and academic support.
- Peer Review: A Peer Review Committee constituted by DAE guides schools to work on six actionable areas aimed at continuous improvement in academic and institutional functioning.
- Alumni and Community: The AEES alumni network ranging from professionals to homemakers provides inspiration and mentorship. Community volunteers and parents also assist in strengthening Foundational Literacy and Numeracy (FLN) skills.
- Rajbhasha: Atomic Energy Central Schools have performed satisfactorily in the inspection conducted by the Parliamentary Committee on Official Language.
- Gyan Samvardhan Lecture Series- Through these, students are exposed to subject experts and inspiring role models.
- Lectures are also conducted for the students, parents and staff members on different topics by alumni, experts and members from the community.

7 Conclusion

The morning assembly at AEES is where "Good Mornings meet Good Morals," reinforcing values like "Less Screen Time, More We-Time", "Padhai bhi, Poshan bhi" and other such themes.

Thus, AEES is more than a school network; it is a living institution of values where excellence is a moving target. By nurturing a balance of knowledge, responsibility, and patriotism, AEES is dedicated to preparing students for Viksit Bharat, ensuring they grow into informed and committed citizens.

These efforts, largely driven by the collective wisdom and innovation of teachers, continue to strengthen the nation's educational goals.

Key Public Health Initiatives for Cancer Control: Training for Cancer Awareness, Detection and Prevention

Gauravi Mishra[1](✉), Harimadhav Viswanathan[2], Vasundhara Kulkarni[1], and Kavita Anand[1]

[1] Preventive Oncology, Centre for Cancer Epidemiology, Tata Memorial Centre, Mumbai, India
gauravi.mishra@tmc.gov.in

[2] Division of Implementation Science and Behavioural Research (DISBR), Centre for Cancer Epidemiology, Tata Memorial Centre, Panvel, India

Abstract. Cancer is a major public health challenge across the world. The rising incidence and mortality highlights the need for a stronger emphasis on prevention and early detection. The Division of Implementation Science and Behavioural Research (DISBR), CCE, ACTREC, Tata Memorial Centre, addresses this through awareness generation, capacity building, community engagement, outreach service delivery and implementation research. The division conducts widespread awareness and screening activities across the state of Maharashtra. It strengthens health systems through training of healthcare providers and the development of standardized Information Education and Communication (IEC) materials. It also conducts multiple implementation research projects to generate evidence for scalable and sustainable cancer prevention models. Through these integrated efforts, the division aims to bridge the evidence-practice gap and improve the uptake of cancer prevention and early detection services.

Keywords: Cancer Prevention · Preventive Oncology · Implementation Science · Cancer Screening · Public Health · Non-Communicable Diseases

1 Introduction

Cancer is one of the leading causes of morbidity and mortality across the globe [1]. As per GLOBOCAN 2022, an estimated 20 million new cancer cases and 9.7 million deaths were reported in 2022. The burden is projected to rise further, with annual new cases expected to exceed 35 million by 2050 [2]. This exponential rise can be attributed to the ageing population, demographic growth, and increased exposure to risk factors such as tobacco, alcohol, obesity and air pollution [1]. A disproportionately high proportion of the cases and deaths due to cancer is expected to be from the low-and-middle-income countries (LMICs) [3].

India has also witnessed a steady increase in cancer incidences over the past few decades, with approximately 1.4 million cases and 0.9 million deaths reported annually [4]. Current estimates suggest that one in nine individuals is likely to develop cancer

G. B. Mohanty and V. S. Prabhudesai (Eds.): DAEC 2026, SPPHY 368, pp. 253–258, 2026.
https://doi.org/10.1007/978-981-92-2948-2_34

during their lifetime [5]. Breast, oral, and cervical cancers remain the most common cancers in the country, while changing epidemiological patterns indicate an increase in the burden of many other cancers like lung, prostate, colorectal and ovarian [4]. This increasing burden underscores the need for strengthened public health action—integrating awareness, early detection and prevention strategies.

The Division of Implementation Science and Behavioural Research (DISBR), Centre for Cancer Epidemiology (CCE), ACTREC, Tata Memorial Centre, advances cancer control through various public health initiatives and implementation research. The division aims to bridge the gap between research and practice by systematically translating evidence into real-world settings. Its initiatives focus on training a wide range of stakeholders in cancer awareness, early detection and prevention, while simultaneously generating evidence to support scalable models under the framework of National Programme for Prevention and Control of Non-Communicable Diseases (NP-NCD). This approach helps identify barriers to the effective uptake and utilisation of existing evidence based interventions and supports the development of strategies to address them. This strengthens the usability, implementation and scalability of proven interventions within routine health systems.

2 Public Health Initiatives

The division delivers its services through multiple initiatives, including outreach service programmes, community-based awareness activities, implementation research projects, training workshops and the development and dissemination of IEC materials.

A major pillar of the division's work is the organization of outreach service programmes. They are organized in the selected localities of Mumbai and Navi Mumbai to deliver comprehensive awareness and screening services. The systemic approach includes—enumeration of the eligible population, followed by structured education sessions and on-site screening for common cancers. Tailored community awareness and screening are organized to cover specific groups. Housing society camps enabled education and screening for urban women residents. NGO collaborations like Family Welfare Agency in chawls helped in reaching to senior citizens, while the DISBR in collaboration with Women's Cancer Initiative (WCI) organized multiple camps across the low socioeconomic wards and other diverse community settings in Mumbai. During such initiatives, special emphasis is placed on high-risk individuals, including those with tobacco and/or alcohol use. This model demonstrates how population-based screening, integrated with targeted community engagement, can effectively reach individuals who may otherwise have limited access to preventive services. It also enables repeated engagement and facilitates effective linkage to screening services.

Aligned with national campaigns, DISBR has implemented women-centric cancer awareness and screening initiatives under the Swasth Nari, Sashakt Parivaar Abhiyaan using trained peer ambassadors. These ambassadors conduct structured sessions at primary health centres, sub-centres, and village settings, promoting awareness on breast, cervical, and other common cancers, along with self-care practices. Rural efforts in Panvel, Palsai (Palghar) and Navjeevan Centre (Murbad) combined education sessions with oral, breast and cervical screening, stressing proximity and documentation for underserved areas.

Building on this approach, the division also leverages non-traditional community platforms for health communication. A two-day programme for 100 women inmates of Central Prison in Byculla, Mumbai, detected 10 breast and 7 oral lesions along with high tobacco/alcohol prevalence. Awareness initiatives conducted in salons, beauty parlours and training institutes engage diverse groups–such as women, youth, staff and clients–through structured sessions and standardized IEC materials. The activities include structured awareness sessions on common cancers, distribution of informational pamphlets, development of educational magazine content and dissemination of video-based IEC materials tailored for use in these settings.

The DISBR identifies workplace as a potential platform for delivering effective awareness, screening and prevention services. The division has organised multiple workplace-based interventions in collaboration with various industries and institutions across the state. Mill workers in industrial settings were given awareness on common cancers along with prevention in the context of occupational and lifestyle exposures. A landmark workplace-based initiative was implemented for police personnel through a dedicated project designed to cover the entire Mumbai Police force. The initiative offered awareness sessions followed by on-site screening for Non-communicable Diseases (NCDs) including common cancers – oral cavity, breast and uterine cervix cancers and structured tobacco cessation counselling. Over time, the project expanded beyond the city police to include Traffic Police, Railway Police, State Reserve Security Force Police (SRPF), Maharashtra Police units in Navi Mumbai, Thane and Mira Bhayandar, as well as the Maharashtra Security Force (MSF), thereby embedding prevention and early detection services across multiple uniformed service. This initiative has become one of the largest of its kind globally, screening over 60,000 individuals and demonstrating the feasibility and scalability of workplace-based cancer control models.

The division organizes programs in collaboration with educational institutions and religious organizations as well. Awareness sessions are organized in schools and colleges in Mumbai and Navi Mumbai area. The sessions in colleges focus on tobacco use and lifestyle-related risk factors, sensitizing young adults to their long-term implications for cancer risk. While the sessions in schools deliver age-appropriate messages on healthy behaviours and early symptom recognition among adolescents. The division has developed picture-based and simple exhibition materials to create awareness, which are displayed in educational institutions for easy understanding and effective communication.

In partnership with religious and community groups, such as Sree Narayana Mandira Samiti and different church/ parish communities, awareness sessions are conducted on prevention and early detection of common cancers, leveraging trusted social networks to improve outreach, acceptability and community engagement. Screening camps are organised either in community halls of these religious groups or interested participants are facilitated and brought to the Preventive Oncology OPD at ACTREC for screening.

The Patient Caregiver Program is a unique initiative targeting attendants and relatives of patients visiting various OPDs at ACTREC. Based on the concept of opportunistic screening, awareness sessions are conducted in high-footfall areas such as outpatient waiting rooms and registration counters. A structured comprehensive awareness session on key topics including cancer warning signs, the health hazards of tobacco and alcohol use, self-breast examination, nutrition and supportive care is delivered by trained health workers. Caregivers are also encouraged to attend the Preventive Oncology OPD to assess their individual risk profiles and undergo routine screening for common cancers.

The research initiatives undertaken by the division are primarily implementation projects aimed at establishing the scalability and sustainability of different cancer screening and prevention strategies. Through close collaboration with public health systems and community settings, these studies evaluate the feasibility, acceptability and effectiveness of delivering evidence-based interventions in real-world settings. Identifying the real-world challenges in implementing cancer screening and developing tailor-made strategies for tackling them in different settings, has remained the key goal for the division.

A major contributor to low screening uptake is the inadequate training of health workers in common cancer screening. To address this, the division is undertaking extensive capacity-building initiatives. In collaboration with the Brihanmumbai Municipal Corporation (BMC), structured training programmes are conducted for healthcare providers, including ASHAs, ANMs and Medical Officers. Through 51 training sessions, a total of 2719 health workers including ASHAs, ANMs, CHVs and Medical Officers have been trained in common cancer screening under this program. Additionally, a hybrid training model-combining online lectures with periodic field-based hands-on training is being implemented in collaboration with the state governments of Bihar and Gujarat to strengthen screening competencies at scale. Around 1500 participants have been trained under this initiative.

The division also organises five unique workshops annually. They are aimed at building capacity among public health professionals, including MD Community Medicine (PSM) residents, MPH graduates, PhD students and research assistants/ associates. Focusing on key domains such as cancer prevention, tobacco cessation and broader public health approaches, they emphasise the development of practical, skill-based competencies among participants. The workshops conducted so far have received a positive response, with participation from over 18 states across India.

Lack of awareness remains a key barrier to the uptake of screening and cancer prevention services. To address this, the division has developed standardized IEC materials—including flip charts, calendar charts, posters and exhibition materials—covering common cancers and the health hazards of tobacco. For widespread dissemination, they are prepared in multiple languages like English, Hindi and Marathi and are made freely accessible under the resource centre tab on the official website of the Centre for Cancer Epidemiology, ACTREC. Their use by different stakeholders can ensure the delivery of standardized, uniform and evidence-based health education sessions across varied settings. The division also undertakes intensified, theme-based activities such as awareness sessions and screening camps aligned with important health days. Special activities are organised on World Cancer Day, Head and Neck Cancer Day, World No Tobacco Day and dedicated breast and cervical cancer awareness months. These time-bound initiatives help amplify outreach, strengthen community awareness and enhance uptake of preventive and screening services (Fig. 1).

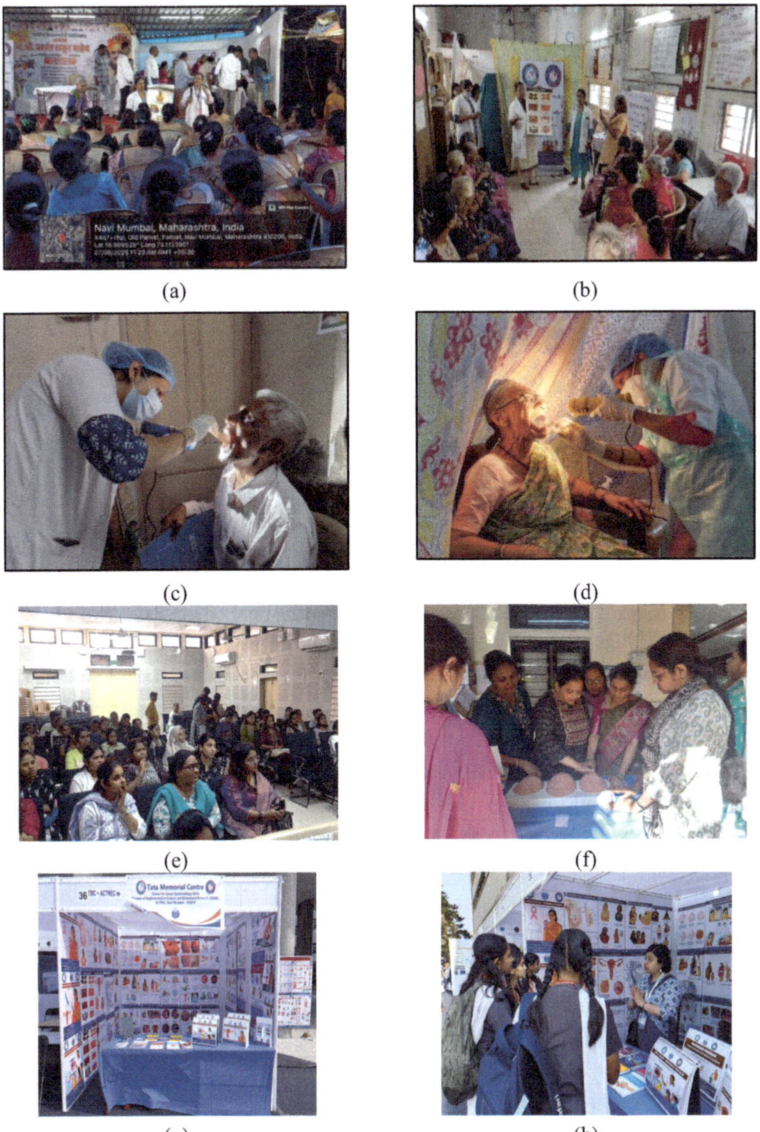

Fig. 1. Different initiatives under the division of implementation science and behavioural research (DISBR), CCE, ACTREC (a), (b) awareness sessions, (c), (d) screening camps, (e), (f) capacity building, and (g), (h) exhibitions

3 Conclusion

The rising cancer burden globally and in India underscores the need to position prevention as the cornerstone to cancer control. Strengthening awareness, organizing screening and early detection remains the key preventive strategies to reduce the cancer-related morbidity and mortality. Although robust evidence already exists for most of these interventions—particularly for common cancers such as oral cavity, breast and uterine cervix cancers—their uptake and coverage remain suboptimal due to contextual, socioeconomic, cultural and health system barriers.

In this context, the role of the Division of Implementation Science and Behavioural Research (DISBR) is critical. The activities of the division enable the identification of context-specific barriers and facilitators and supports adoption, integration and scale-up of evidence based strategies in real-world settings. The public health initiatives outlined above demonstrate how DISBR operationalises its mandate through ongoing activities in awareness generation, capacity building, service delivery and research.

4 Future Directions

The primary focus will be on scaling up and integrating evidence-based cancer prevention and screening strategies within existing health systems, particularly under NP-NCD framework. This would ensure wider and equitable coverage of screening and early detection services, thereby strengthening overall cancer prevention efforts. Strong emphasis must be placed on awareness generation among public and capacity building of healthcare providers. Community engagement should be embedded across all activities to ensure acceptability, ownership and better uptake of interventions.

A special emphasis will be placed on reaching high-risk and underserved populations, where both disease burden and lack of awareness are higher. Newer technologies like digital health platforms and artificial intelligence will be used for creating awareness and improving screening uptake. There is a need to shift research priorities towards implementation research to address the real-world challenges and bridge the research-action gap at the ground level. Finally, multi-sectoral and inter-departmental collaborations will be established at every level for effective and sustainable cancer control.

References

1. World Health Organization: Cancer [Internet], Geneva: WHO; [Cited 2026 Apr 18].. Available from: https://www.who.int/news-room/fact-sheets/detail/cancer
2. International Agency for Research on Cancer: Global Cancer Observatory: Cancer Today. GLOBOCAN 2022: All cancers fact sheet [Internet]. Lyon: IARC; 2024 [Cited 2026 Apr 18]. Available from: https://gco.iarc.who.int/media/globocan/factsheets/populations/900-world-fact-sheet.pdf
3. Shah, S.C., Kayamba, V., Jr Peek, R.M., D.: Heimburger cancer control in low- and middle-income countries: is it time to consider screening? J. Glob. Oncologia. **5**, 1–8 (2019). https://doi.org/10.1200/JGO.18.00200. PMID: 30908147; PMCID: PMC6452918
4. International Agency for Research on Cancer: India fact sheet. In: Global Cancer Observatory. Lyon: IARC (2024). Available from: https://gco.iarc.who.int/media/globocan/factsheets/populations/356-india-fact-sheet.pdf. Accessed 18 Apr 2026
5. Sathishkumar, K., Chaturvedi, M., Das, P., Stephen, S., P.: Mathur cancer incidence estimates for 2022 & projection for 2025: Result from National Cancer Registry Programme, India. Indian J. Med. Res. **156**(4&5), 598–607 (2022). https://doi.org/10.4103/ijmr.ijmr_1821_22. PMID: 36510887; PMCID: PMC10231735

Author Index

G. B. Mohanty and V. S. Prabhudesai (Eds.): DAEC 2026, SPPHY 368, pp. 259–260, 2026.
https://doi.org/10.1007/978-981-92-2948-2